# Canine
# aggression

## Rehabilitating an aggressive dog with kindness and compassion

**Tracey McLennan**
MA (Hons) BSc (Hons)
Foreword by Sarah Fisher

Hubble & Hattie

The Hubble & Hattie imprint was launched in 2009, and is named in memory of two very special Westie sisters owned by Veloce's proprietors. Since the first book, many more have been added, all with the same underlying objective: to be of real benefit to the species they cover; at the same time promoting compassion, understanding and respect between all animals (including human ones!) All Hubble & Hattie publications offer ethical, high quality content and presentation, plus great value for money.

## More great books from Hubble & Hattie –

# www.hubbleandhattie.com

First published April 2018 by Veloce Publishing Limited, Veloce House, Parkway Farm Business Park, Middle Farm Way, Poundbury, Dorchester, Dorset, DT1 3AR, England. Tel 01305 260068/Fax 01305 250479/email info@hubbleandhattie.com/web www.hubbleandhattie.com ISBN: 978-1-787110-79-3 UPC: 6-36847-01079-9 ©Tracey McLennan & Veloce Publishing Ltd 2018. All rights reserved. With the exception of quoting brief passages for the purpose of review, no part of this publication may be recorded, reproduced or transmitted by any means, including photocopying, without the written permission of Veloce Publishing Ltd. Throughout this book logos, model names and designations, etc, have been used for the purposes of identification, illustration and decoration. Such names are the property of the trademark holder as this is not an official publication.

Readers with ideas for books about animals, or animal-related topics, are invited to write to the publisher of Veloce Publishing at the above address. British Library Cataloguing in Publication Data – A catalogue record for this book is available from the British Library. Typesetting, design and page make-up all by Veloce Publishing Ltd on Apple Mac. Printed in India by Replika Press.

# Contents

# Dedication, Acknowledgements & Foreword

## Dedication

This book is dedicated to the men, women and children of the Glencoe Ski Club, for teaching me to thrive on adversity.

## Acknowledgements

I owe a debt of thanks to the many people who helped Calgacus and I to recover and find joy in life again. I can't name them, but in these I include all the kind strangers who let Calgacus and I walk close to their calm dogs; those who let us follow them around the park so we could practise, and those who came and admired and patted Calgacus and Katie. That sort of encouragement lifted many a dark day for me.

And there were, of course, the amazing people who actively helped me with Calgacus by walking with us, or endlessly talking to me about him, or by teaching me the skills that I needed. So, to Louise Williamson, Marie Miller, Jane Pelly, Jo Law, Tina Constance, Alison Sinclair, Gil Vick, Freya Kennedy, Jeanette Fyfe, Graham Cameron, Sarah Fisher, Yolande Bosman, Helen Maunder, Fiona Forest, Penel Malby, Laura McAuliffe, Selina Clarke, Janet Finlay, Robyn Hood, Jane Fawcett, Gordon Fawcett, and Sally Jeffrey, thank you. I owe all of you an enormous debt for the various ways that you helped me through one of the most difficult times in my life.

Then there were the special dogs who helped out, providing their skills and care in helping Calgacus re-learn how much fun it is to play with his own kind. To Maisie, Ole, Sally, Bracken, Inis, Paddy, Willow, Bilbo, Dobbie, QT, Harvey, Louis, Sid, Skye, Hattie, Morris, Grace, Barley, Lola, Saxon, Pickle, Luka, Beama, Rufus, Mirri, Jake, Florence, Grace, Orsa, Corrie, Willow, Oz, and Kaya – thank you.

Sarah Galloway and Jane Vegglis: I think that you both listened to more than your fair share of my dog-related woes at work. Thank you, both.

Finding things to do once Calgacus and I felt better about other dogs, learning to cope without him, and writing this book have similarly been collaborative efforts made better by all the wonderful people who have provided help and support.

Even with all the miles that separate us, Sharon Donohoe, Penny Downing, Angie Anderson and Julia Dobbs, your help and support through college and beyond made things all so much easier, and more fun.

Rebecca Leonardi, Freya Kennedy, and Jen Gates – you have all given me wonderful voluntary work, and helped fill the gap that opened up when I finished studying and found myself suddenly with spare time on my hands.

For all the time spent outside in various places practising for photos, thank you,

4

Rebecca Leonardi and Reka Toth, as well as our brilliant guides and teachers – Cuillin, Roxy, Ellie, Jet, Bonnie, Mojo, and Karma.

Lynne Mejury, Freya Kennedy, Louise Downing, Sharon Donohoe, Sarah Fisher, and Penel Malby: your words of support and suggestions for improvements of an early draft of this book made continuing work both easier and more productive. You are all fantastic.

In terms of helping me gain more writing experience and getting on with producing the book, I owe thanks to Denny Writers. Thanks must also go to Rowena Murray for giving me valuable space on her writing retreats, and to all my fellow writers on those retreats who were gracious and kind as they worked hard on serious academic works whilst I sat among them writing a book about a dog.

Of course, thanks are due to my lovely family, without whom I wouldn't be the person I am now, and who are endlessly patient about my desire to be doing things with dogs instead of going skiing.

Many thanks, as well, to my wonderful partner, Gordon Asher, who has encouraged me at every stage of the process of writing this book, and who has learned enough about dogs to enable him to help with editing an early draft of it.

## Foreword by Sarah Fisher

I first met Tracey and the glorious Calgacus when they came to Tilley Farm to explore Tellington TTouch; both became much valued friends. I was thrilled when Tracey asked me to read her book, as although I knew a little of the struggles that she experienced as Calgacus began to mature, I didn't know the full details of their time together.

This beautifully-written book is an honest, moving and powerful account of the dreams and hopes of someone seeking to share their life with a dog for the first time. It highlights the challenges faced by a responsible and committed novice dog guardian as reality sets in, and charts the progress of Tracey's own education as she navigates her way through a myriad of beliefs and myth-understandings, seeking support for both herself and her mighty canine companion.

As well as being a compelling read for those starting out on the path towards building a successful and rewarding relationship with a dog, or those struggling with a dog already in their care, *Canine aggression* serves as a reminder to all who work in the field of canine training and behaviour that we have a duty to support the human client with the same compassion and understanding as we do their dog. Dog guardianship can be a confusing and utterly overwhelming experience for many, and some of those to whom canine guardians turn for advice and support can leave them feeling more isolated, judged and insecure, albeit not intentionally. We need to be mindful that there is always a back story, whether or not we are aware of it, and remember that 'learning to teach dogs isn't a simple thing,' as Tracey so wisely writes.

From a well-intentioned dog lover who quickly found herself out of her depth, Tracey has become one of the most gifted, thoughtful and compassionate canine care-givers and educators I know. She continues to help and enrich many lives, and I will be eternally grateful to Calgacus for bringing her into mine. Calgacus will always hold a special place in my heart for many reasons, and by telling their story, Tracey can share his magic with those who did not have the good fortune to know this wonderful dog.

# 1 It started ...

... with a puppy. The first one I'd ever lived with. I grew up in a family without dogs, yet dogs were all I could think of. When life allowed I found Calgacus – a striped brindle Bull Mastiff puppy – and I was sure that having him in my life would make it complete. As the years have passed, I have become more and more convinced that I – and the rest of humanity – needs dogs: to remind us to be humane, to help us through our darkest times, and to help us have fun.

I want to add my voice to the many others talking and writing about lives transformed through relationships with dogs. I want to share the story of how my eleven-year relationship with a dog changed me. The catalyst for this change was traumatic: the puppy I'd dreamed of for so long grew up, and, before he was two years old, was behaving aggressively toward other dogs. The changes for me began there – in blood, despair, and confusion – with a desire to 'fix' my dog – which changed as he and I walked together, learning as we went, until he became friend, mentor, and playmate to many other dogs. By the end he wasn't 'fixed' because he'd never been broken. He was changed, and I was changed just as much as he.

As we learned, I noticed that I was not alone in searching for ways to reduce aggressive behaviour in dogs. My friends and colleagues who work and live with dogs are similarly interested in the topic. It is usual for any new technique for aggression to result in vast tracts of discussion in writing and video form, as well as people travelling long distances, fuelled by rich, dark, coffee or energy drinks, to learn from the gurus who have the answers they seek. I've heard the search for ways to work with aggression be described as similar to looking for the Holy Grail: obsessive, long-term, fraught with difficulties, and with uncertain results.

During my own exploration of this world, some lessons surfaced again and again in relation to how I related to Calgacus – and any dog I came into contact with, as well as other people. The first of these was that being an authoritative leader of dogs (or people) who obey without question is not useful or kind. Instead, a questioning attitude toward life, and the ability to enter into a dialogue, proved of much greater use.

This particular lesson – luckily for me – was something that my work outside of dogs helped with. I have been a computer programmer since 1997, and the unquestioning compliance of computers has demonstrated to me how bad we are at thinking through the implications of our instructions. I've spent a great deal of time – sometimes into the early hours – fixing the unfortunate consequences of poorly thought out computer instructions. A thinking, questioning companion is far easier, and much more enjoyable and rewarding, to work with than one who is blindly obedient.

Another valuable lesson was learnt whilst seeking ways to manage an aggressive dog. Change can be stressful and unsettling, even when it is desperately wanted, and when I began to get the results I wanted – a dog who was friendly toward other dogs – I was terrified. I'd become used to avoiding other dogs, walking in quiet places, keeping Calgacus safe. Taking steps to let him interact with dogs again was as stomach-churning as standing on the edge of a cliff and staring into a deep abyss.

Teaching and learning taught me that what had previously appeared familiar, was actually of greater expanse and more colourful than I could have imagined. My teaching style had been the transfer of knowledge from one being to another, and learning as receiving that knowledge: a monochrome, emotionless, one-direction process. As Calgacus and I worked together, I began to realise that there was much more to it; that knowledge flows from teacher to learner and back again, looping and finding new pathways until it is impossible to say who has learned the most and who has taught the most. I discovered that *how* teaching is done matters; paying attention to tiny detail is important, and taking time to rest is also important. The process is rich with emotion.

The last lesson for me was that I needed to reach out for help; to realise that I was not alone in what I was struggling with. I found people all around me who were willing to offer help and support, if I just looked for them and reached out. Living with a dog who can be aggressive toward their own kind can be an isolating experience, filled with the stress of dealing with problems that other people are blissfully unaware of. Each walk required careful planning in order to avoid conflict, and was taken at times when few other dogs might be around. Even then, the shame and dismay I felt when Calgacus erupted in a barking frenzy at somebody else's friendly dog were hard to bear. A natural response to this is to withdraw, avoiding people who just don't understand what it is like to have a difficult dog, yet one of the greatest discoveries for me was just how kind, helpful and supportive most people actually are. The journey that Calgacus and I made would have been different, and so much harder without the people – and their dogs – who helped us along the way.

My hope is that this book will provide a glimmer of hope, and be a source of comfort, support and ideas for people in situations like mine. People who love their dog, but who are feeling lost and out of their depth, because the relationship is causing them stress and anxiety, rather than filling them with joy.

# 2 Choices, choices

"Would you do it again?"

"Would you choose a Bull Mastiff as a first dog if you could do it all over again?"

"Would you recommend that breed for somebody who has never cared for a dog before?"

"What would make somebody with no experience with dogs want to live with a Bull Mastiff?"

I used to get asked these questions often, and my answers remain the same –

"No," to the first three and "I didn't know enough about dogs to understand how difficult it would be," to the last one.

I hadn't blindly chosen to look for a Bull Mastiff. I'd done my research; I'd read all the books I could find on numerous breeds of dogs, and then read the popular books about dog behaviour, and how best to teach dogs at that time. The books about behaviour pretty much agreed that a dog is a dog, and that, as long as the person caring for them is consistent and a good leader, any dog at all will behave well. Turns out, I read the books written by well-known individuals, unaware that there were others that might have different (and better) ideas. I wasn't reading the right books. Obviously.

Back then I didn't know what I know now. Now, I'd say that even if one dog is pretty much the same as any other dog – which I'd argue that they aren't, in any case – learning to train dogs is far from simple. It takes time to acquire the mechanical skills needed, and more time to hone and develop them to the point where it is possible to have a good teaching/learning relationship with your dog. For those living with a dog for the first time, the situation can be even trickier. They have a huge amount of learning ahead of them, and, to make matters more difficult, as the dog training and behaviour industry here in the UK is unregulated, it is very easy to become involved with a trainer who actually makes things worse ...

To tackle challenges like these is harder still when the puppy in question is large. At ten weeks of age, Calgacus weighed ten kilos, and he got bigger – fast: by ten months, the ten kilos had become forty, and his energy and strength seemed ten times greater. He was an adolescent: a challenging time of life for dogs and people alike. Like truculent teenagers, dogs of this age often stop co-operating with any requests made of them, and will sometimes use strength or cunning to get what they want. Many of them, having learned not to chew on furniture, wires, shoes, books and mobile phones, begin doing so again as they reach adolescence, and those living with them often despair, not understanding why their dog – who had been getting to be so well behaved – has turned into such a nightmare.

Then there is the guarding behaviour that Bull Mastiffs are prone to. Bred to be guardians, they were the dog of choice for gamekeepers, and often worked in the darkest parts of the night, in the wild woods and moors, slipping quietly through the dark in search of poachers, holding them in place until they could be arrested.

Modern Bull Mastiffs are, thankfully, not the objects of fear that their ancestors must have been, though there is usually still a point in their adolescence where they begin to become suspicious, and sometimes of things that, previously, had not bothered them. A lone person walking in the purple light of dusk on the other side of a field can cause concern, for example, and a young Bull Mastiff may react by barking madly and pulling toward the person, or turning and trying to run from them. Either scenario isn't particularly easy to deal with. Accounts of patting road signs, and pretending to feed them cheese as a way of showing that they are 'friends' are common amongst people who live with Bull Mastiffs.

I know that my introduction to having dogs would have been easier and happier – and would not have entailed a complete lifestyle change – had my first dog been physically smaller – a Spaniel of some kind, perhaps, or one of the small, fluffy companion breeds.

I cannot claim that my choice of dog breed was entirely down to the books I read. They were a big influence, yes, but my past experiences and world view wove into the picture, too. Back then, things looked good for me. I had an interesting, well-paid, professional job. A loving and similarly employed partner. A close family; friends; a large and beautiful home, and my own car, plus lots of fun social time outside of work. A good life.

But I wasn't okay. I felt a desire to withdraw from the world, and did so, as much as I could. I let close friendships drift, resulting in the people I loved moving on with their lives without me. My tendency to be a daytime napper grew, until, at weekends, I regularly found myself getting up, eating breakfast, and then returning to bed until lunchtime. Unless I'd been on a night out, these days were preceded by evenings spent sleeping in front of the TV. I amazed myself with my capacity and desire for sleep.

My habit of waking early in the morning and having that time to myself hasn't changed, but, prior to having Calgacus, I found it impossible to stay awake through the day and evening unless I absolutely had to. More and more often, I chose to avoid doing things that would keep me awake. I stopped skiing during winter weekends, and so spent less time with my oldest friends. I found excuses to put off making plans with other friends, saying things like "We haven't seen each other for ages. We must make plans to get together. Yes. We must make plans. Well, then, what I'll do is ring you some time and we can make a plan to get

# Canine aggression

together," which made it sound as if I wanted to meet them, and I did want to, I just didn't want to be tied to a plan. Plans got in the way of sleeping ...

Time spent with my family was kept short, in order to allow me to sleep more. Holidays with my family became hard work and exhausting; not the source of enjoyment that others experienced. Although I lived with my partner, and had friends and family I got on well with, I felt lonely. Of course I did. Prioritising sleep over social time made it impossible to connect with others in a meaningful way.

The thing is that it didn't *feel* like a choice. I was so tired, and tried to deal with this by sleeping more, as I'd been doing. My friends were and are busy people with their own lives, and so, when I wouldn't make plans but otherwise seemed fine, nobody had reason to be concerned. They simply got on with their lives, and gradually I became more and more isolated.

I daydreamed about living in a remote cottage, far from others, where I could hide. Dogs were caught up in my thinking. I loved them – had done since I'd been a child – but I'd never lived with one: my social activities and hobbies left no time for dogs. My understanding, then, of having a dog was that they prevented families from doing other things: a tie and a barrier to joining in with social events. It was no surprise, then, that my thoughts turned to dogs as I sought solitude.

Maybe depression had me in its grasp? I just don't know. I felt fine – if a bit tired – and told myself that I was independent; someone who liked their own company, and who loved to sleep. Both true. I had been taking naps for a long time, I reassured myself (also true). Maybe if my changed lifestyle had interfered with my work I might have considered it a problem, but as long as I could go to work when I needed to, and maintain a small amount of social contact with friends and family, I felt okay about it. Not great, for sure, but better than lots of people, and certainly not struggling.

Anxiety had a part to play, too. My family describe me as a worrier, and their stories about how this has impacted on my behaviour are numerous. As a toddler I sat down on pavement edges to reach the road rather than stepping off the kerb and risk falling over. At three, left in front of the TV while Mum cared for my younger brother, I watched a news bulletin about an old satellite with a deteriorating orbit, due to fall to the earth at some point the following week. Poor Mum subsequently had to coax and cajole me to leave the house for days afterward, convinced, as I was, that we would be squashed under mounds of metal falling from the sky.

I would also scream so loudly every time Mum tried to go into the local wool shop that she was forced to leave me outside, and it wasn't until I was old enough to talk that I explained that my fear centred around the shop's decapitated head dummies modelling knitted hats and scarves, and resting on shelves rather than shoulders.

As a teenager, I lay awake for hours every night, muscles rigid with fear, sipping in air; too afraid to take normal breaths in case the sound of my breathing prevented me hearing the monster I was sure was creeping through the house.

It is safe to say that anxiety has always been a part of my make-up, and, if anything, my imagination has become even more inventive in this respect as I've gotten older.

Friends sometimes ask me "What's the worst that can happen?" and I'll answer with something like "Light could flash off my watch face when I move, causing a passing driver to be blinded so that they mount the kerb and kill an innocent person on the pavement."

This is usually met with a wide-eyed stare, and a fervent "I'm glad I don't have to live in your head!"

As an adult with my own home and job, I was frightened to walk around alone for fear of being attacked. Isolated, rural places scared me the most, as I imagined people with malicious intent lurking behind every tree, the isolation providing them with the ideal opportunity to attack me. My home is surrounded by beautiful countryside, but, in the days before I had dogs, I was unable to enjoy it.

This unhappy state of affairs – my fears and desire for isolation; the books I read – coalesced into a perfect storm that led me to seek out a Bull Mastiff puppy. The books I read – picturing sleek, glossy images of well behaved dogs on the covers – emphasised the importance of obedience in dogs and leadership in people, issuing commands that dogs obediently obeyed. By following a few simple rules, the books claimed, *any* dog would become a loving and obedient companion who would constantly look to their carer for guidance. They made it all sound so simple; so easy.

The reality is that a dog whose ancestors were bred and trained for a task such as Calgacus' were, are likely to be very different in size, shape and mental attitude than one bred for the life of a lapdog, or to retrieve, or safely guide a blind person, or listen as a child reads a book. But I didn't know this then, and believed what the books said. If I *had* known – if the worried and lonely person I was then could have seen what lay ahead, I may well have chosen differently ...

# 3 Looking back

I am unable to recognise the me who existed before Calgacus came into my life. I can remember what I believed about the world and about other people, but the memories are dream-like – as though from a film I once watched. As much as I know she once *was* me, I feel totally disconnected from that person.

Back then, before dogs, I strongly identified with the philosopher Thomas Hobbes. In his book *Leviathan*, he describes a humanity filled with darkness. To him, people are frightening individuals, who wear a civilised facade to cover their aggressive and selfish core, and he argued that strong laws, backed up by harsh punishments, are necessary to force people to hold their brutal urges in check. Without these regulations, he envisaged a bleak world in which nobody would even grow food, since to do so would result in it being stolen. Instead, people would spend their time fighting and killing each other just to stay alive – living short, miserable lives.

I also believed that anybody who harmed another person did so deliberately, and in full knowledge of their actions. I looked at the world, saw the harsh ways that we treat each other, and took this as proof that Thomas Hobbes spoke the truth. I agreed with him, and felt that we needed strict laws, and a society that consisted of leaders who made and enforced the laws, and followers who obeyed them. How else could our violent tendencies be kept in check?

Toward the end of the 1990s, I began researching dogs . To anybody reading this who has knowledge of dog training and behaviour, it will come as no surprise that I could easily find books that resonated with my bleak beliefs. One of the first books I read advised that dogs need strong leaders; a structure; rules, and stringent enforcement of them. It listed a few of the rules, and gave simple instructions about how to ensure they were followed.

Reading that book evoked in me memories of days curled deep into the warm brown leather of a chesterfield armchair in the reading room in the Glasgow University Union, a  sturdy, white porcelain mug full of coffee – strong and bitter enough to see students through nights of studying – in hand as I read Thomas Hobbes. The words this time were about dogs rather than people, but they resonated so strongly that I nodded as I read, happy that this was so familiar. I needed no convincing that, without strict guidelines, dogs would attempt to improve their position in the household hierarchy and take control, becoming aggressive, or anxious or destructive as they did so. The way to prevent these problems, the book claimed, was by mimicking the behaviour of leaders in a wolf pack: eating your meal before feeding your dog; always walking in front of the dog; being the only one to initiate social contact.

Just as within the pages of *Leviathan*, there was a lot of darkness in the book, with lurid descriptions of what would happen if its advice was not followed: children being bitten, and dogs fighting with canine companions, in each case citing the cause as a lack of leadership and hierarchy. Amazingly, a simple act of omission – forgetting, for a while, to behave in this rigid, dictatorial way – was enough for dogs to begin fighting each other for control of the household, it was claimed!

The book was new and glossy, and featured prominently on bookshop shelves. I was enthusiastic about it: not only did what it say make perfect sense to me – since it fitted so closely with my view of the world – but the prominence it was afforded in pet shops and bookshops convinced me that experts in dog behaviour also thought that it was good. As I was to later learn, this idea of behaving like the leader of a wolf pack is commonly known as the 'dominance theory,' and the advice in the book is known as a 'rank reduction programme' (RRP) – intended to ensure dogs understand and accept their lower status in the household hierarchy. Decades-old research into wolf behaviour suggested that packs operate in a strictly hierarchical way, with the alpha pair at the top. This work was read by some dog enthusiasts who, noting, that dogs and wolves are similar species, thought that this must apply to dogs, too. They adapted the work of the scientists, and noticed that a family dog would often become calmer and easier to live with. Popular in the 1980s, the dominance theory enjoyed a resurgence in the late 1990s when I was busy researching dogs, and remains popular still, to a certain extent – in part, due to continued support by a few high-profile authors and presenters of television programmes.

In time, I was to discover that other researchers and practitioners – including some who were once enthusiastic about the dominance theory – have become sceptical of it. I have heard it described, by a previous enthusiast, as a method that 'sucks the joy out of a dog's life,' making the animal apparently calmer through depression and lethargy, rather than natural behaviour. One of the prominent wolf researchers responsible for espousing the dominance theory has now had second thoughts, suggesting in their more recent work that it is a much more complicated issue than previously believed.

Digging deeper, the assumption that domestic dogs need a strict social hierarchy because they are like wolves does begin to look ever-more doubtful. In research and writing, social hierarchies are attributed to many species; I've read articles on this subject about cows, degus, and chickens, for example: indeed, the term 'pecking order' is often used to describe social hierarchies, and derives from studies about chicken behaviour. One of the scientists in the film *Project Nim* (https://en.wikipedia.org/wiki/Nim_Chimpsky ) talks about Nim – a chimpanzee – being dominant, and how she would bite him on the ear to show him that she was in charge. Those who care for parrots are advised sometimes to not let their pets stand on their heads because the parrot may think he or she is in charge. Certain behaviour of a Giant Pacific octopus toward a researcher has even been

# Canine aggression

described as a dominance display. Giant Pacific octopuses are a solitary species, only coming together toward the end of their lives when their bodies tell them it is time to reproduce, so I cannot imagine what need they would have for a social hierarchy. I've yet to hear anybody claim that cows, degus, chickens, chimpanzees, Giant Pacific octopuses and parrots are wolves by any other name, and although I have little knowledge of them, I confess to being sceptical that all of these species lead such rigid social lives. In recent years, I've found myself wondering whether the tendency to see hierarchy everywhere says more about the attitude of the academics and others who adhere to these theories than it does about the animals ...

I'd like to be able to claim that I discovered all of this information whilst researching dogs, but I didn't. I'd found the dominance theory and I liked it: it fitted my world view, and provided me with the reassurance that, no matter what kind of dog I got, he or she would fall in line if I just behaved like an alpha wolf. I did not see any need to look further.

Once Calgacus moved in, however, doubts about this began to creep in. I appreciate that it is jumping ahead to talk about Calgacus as a puppy at this point, but it feels like this is the right place to talk about him, since one of the first things he taught me was to question my view of dogs as aggressive social climbers. Calgacus arrived in my life – confident, self-contained, and self-assured – and made me question all that I'd read and taken as gospel. I was in love with him, and I wanted him to love me back. I didn't want to make a hungry puppy wait for his food while I ate first – it seemed like such a mean thing to do, and especially to such a young pup who'd only recently arrived. Nor did I want to make him walk behind me all the time, and I certainly didn't want to have to go out of the door first when I was rushing to get him outside for the toilet when he was learning that outside was the place to do this. I wanted him to sit on the couch with me, and I liked it when he was the one who came to me and initiated contact.

We sat together on the couch for hours – a respectful gap between us – because, even as a young puppy, Calgacus knew his own mind, and personal space was important to him. He did not like being cuddled, so I would reach across to stroke his soft, baby fur, marvel at the delicate spongy texture of his paws, and whisper that I loved him. In those quiet, perfect shared moments, I could not imagine that this puppy would plot to take over the household. Calgacus, without being able to speak a word, made me dismiss everything I'd read about dominance theory.

The simple reality of him was enough to disprove all of it.

# 4 A friendship begun

I found Calgacus via a notice in a pet shop. Even now, all these years later, thinking about it, I am elated: elevated into awareness on a wave of enthusiasm.

"Bull Mastiff puppies available. Ready in two weeks," the sign read. Below the wording was a local phone number. Bull Mastiffs nearby and ready for new homes soon; during a period I'd be able to take time off work. I rushed home, driving quickly, excited while telling myself not to be. The chances were that this wouldn't work out: that I'd find the number belonged to somebody uncaring, seeking only to make money out of their dogs; somebody I wouldn't be happy about getting a puppy from.

I phoned as soon as I got home.

"Hi. I'm calling about the puppies you have advertised."

"Do you want a boy or a girl? We have both. There are eight puppies in the litter. Four boys and four girls. What made you think about a Bull Mastiff? Do you have any other dogs? Do you have a garden? Do you work? Have you ever had a dog before?" The voice at the other end of the phone belonged to a young woman. She sounded bright, happy, and enthusiastic.

The questions went on and on, and, in the end, the young woman – Susan – and I spoke for over an hour. The love she felt for Lola, the mother of the puppies, rang in her voice every time Lola was mentioned, making her words sparkle.

That night I lay awake for hours, head filled with thoughts of a puppy, and hope that the visit the following afternoon would go well. The time passed like treacle poured from the tin, but pass it did, and finally we were there – my then partner and I – standing in a small, concrete yard outside a white, PVC door, waiting for it to be opened.

We were greeted by an enormous, waggy-tailed dog. Her short, smooth fur was a rich, russet colour, and had black stripes running through it. She sniffed at us with a broad, square snout – the fur covering it was inky-black and felt soft, like the sensitive nose of a horse. Dark umber eyes gleamed when she looked at us. Lola was gentle, and moved slowly as she snuffled our hands. She stretched out her neck to encourage us to scratch her chin, and that first touch was magical. She felt like living velvet under my hands, and I had a moment of disbelief that I might have the chance to live with a dog like her. Once she'd spent time with us, she took herself off to lay quietly on the floor.

We were settled into the cosy living room – a smallish space with soft carpets, and the sort of deep, comfortable sofas that it is easy to curl up in and hard to get out of. A large bottle of carpet cleaner stood in the corner, providing

# Canine aggression

evidence that the puppies regularly spent time in the room. We were given mugs of tea that tasted strongly and reassuringly of tannin.

Once Susan and Neil were content that we had been properly welcomed into their home, the four male puppies were brought in. Just those four, at my request. I felt strongly, and for no particular reason that I could put my finger on, that I wanted to give a home to a male puppy. Plumply, they wandered around the floor – two reddish-brown puppies, a pretty cream one, and, finally, a puppy with the same brindle colouring as Lola. The cream puppy sat quietly, a bit apart from the others. He was friendly, gentle, and calm when called over for attention. My partner liked the way that this calm, cream puppy interacted with us.

"This one is so sweet. Look at how gentle he is, and look at how he is watching what's going on without hiding or being pushy. I really like him."

Probably, my partner was right, and the cream puppy would have been a good choice, but I was losing my heart to the brindle one who was a whirlwind of bright stripes and chubby little legs. He approached, looked at me, and then grabbed me by the trouser leg, his sharp teeth scraping my skin. I winced from the sting of it as he took hold. He braced himself, pulling back so that the fabric tented outward, and then he hung on, growling a high-pitched, puppy growl. I had no idea how to handle this sort of interaction – if I'm truthful, it scared me a little. This puppy had a wildness about him.

I picked him up, hoping to calm him and have a cuddle with him, but he wriggled furiously and struggled to get back to the floor. When I put him down and he went away to wrestle with one of the other puppies, I watched him, and my heart swelled with love. I had come to this meeting expecting to see some cute puppies; hopefully, one who I would feel a bond with. I hadn't expected this rush of joy, fear, love and hope, combined in a confusing, awakening burst of emotion.

"I really like him," I said, pointing to the brindle puppy. To express what I was really feeling would have felt silly, so this was all I could say.

Susan's eyes gleamed; she smiled at Neil. "That's Calgacus. He was named after the first historical Scot ever recorded: a Celtic warrior who fought the Roman army during its occupation of Britain. Calgacus was the first-born: he's the biggest puppy and we love him. If we were going to keep one of the puppies, it would be him."

Neil grinned, nodding, and added, "We'd love to keep him but we just can't. We have Lola to look after, and a couple of other dogs, too. We just don't have the time or money for another dog."

Their words sealed it for me. I'd felt that the brindle puppy was special and they had confirmed it. On the journey home, I expressed enthusiasm about Calgacus until my partner agreed that he was the one we should take home with us in a few weeks, when he was old enough.

The attention to detail and love for their puppies that Susan and Neil had bestowed on them made all the difference in Calgacus' early life, and set him up well. The couple had a constant stream of visitors and friends to their home, who played with, stroked, and held the puppies, enabling Calgacus to grow up knowing how much fun people are, and being keen to make new friends. He learned, too, that trips out in the car were fun, and usually culminated in visits to offices, train stations, and shops. In the mornings, when the post was delivered, the postie was greeted by a happy Lola, who was delighted to be patted and fed biscuits. Calgacus and his siblings enjoyed similar attention, and he grew up with a lifelong love of posties.

I have no doubt that the high quality of this early upbringing and socialisation made a massive, positive difference to Calgacus. He really was set up well with a mother who was happy, enjoyed company, and was very much loved. On top of that, he spent those first, crucial weeks of life in a home where he was loved and allowed to learn about the outside world safely. Without this foundation, I think things might have been different for him – particularly because he was never going to be an easy, uncomplicated companion.

Even before he was weaned, Calgacus was showing signs of being a handful. Susan and Neil talked often about his feeding habits, to the extent that these filtered from the narrative of their lives into mine. The story we heard most often was that, when feeding, Calgacus would take the nipple closest to his mother's front or hind legs, and tug on it until he could position himself between her legs and suckle from there, effectively hiding himself from view.

After the puppies had been fed, Neil and Susan would take them from Lola so that she could go outside and wander around. Concealed Calgacus would sometimes be missed, and remain where he was for a few more minutes, filling his stomach with warm, comforting milk. The mistake would be noticed as the puppies were counted and found to be one short, and a more thorough search revealing Calgacus, snug between Lola's legs, still suckling.

The weeks passed and the puppies grew stronger, gained independence, and developed personalities. Calgacus showed signs of determination, and of having strong opinions about the world. We visited one afternoon when he was five weeks old. The soft carpet showed some signs of wear from the puppies, who played and sometimes toileted on it. This time, we drank dark and bitter coffee, the mugs handed to us by Susan, whose eyes were shadowed by tiredness.

"This is the first time I've sat down all day." She said. "I can't remember the last time I ate a complete meal." Neil nodded. "I can't believe how tiring it is having the puppies – we're on the go constantly."

They sat in those comfortable chairs, exhaustion showing in the way their shoulders slumped. We sipped coffee and watched Calgacus playing with one of his slightly smaller litter mates. Bouncing happily around the room, he repeatedly knocked over the smaller, darker brindle puppy, clumsily smacking his paws on her body, and rolling on top of her.

She struggled to her feet, youth making her unsteady, and padded across the carpet away from Calgacus. Her tail was tucked under her body and, even to an inexperienced watcher like me, she looked like she wanted the game to stop.

# Canine aggression

With a wagging tail and a bounce, Calgacus caught up to her and pushed her over again. He looked delighted to me, as if he was really enjoying his game.

With a sigh, Lola got to her feet. She stared intently at Calgacus, and I could hear a growl coming from deep within her massive body. Calgacus ignored her – too engrossed in knocking over his smaller sister again, and rolling her around on the ground. Lola padded closer – her legs massive, solid and steady next to those of her puppies. Moving to where Calgacus was, she put her head down toward him. His whole body could have fitted easily inside her mouth. She stared at him, growling louder this time.

Calgacus waddled around her, and knocked over his sister yet again, bouncing over and on top of her. With her nose, Lola pushed him back providing a barrier between him and his sister, and then further back again. This time when he tried to go around her, she put her mouth around the back of his neck, holding him in place. Her body was stiff, and, to my untrained eye, she looked annoyed.

Calgacus stood still. Nothing more than that. He didn't yelp or cower or try to get away from his mum; just stood and waited. Within seconds, Lola moved away, her point made. He was being too rough; he had to leave his sister alone.

Freed from constraint, Calgacus ran – his chubby, puppy legs struggling to provide the speed he wanted – to his sister and pushed his shoulder into her to knock her over onto her back. This time Lola was less gentle. Barking sharply, she moved toward Calgacus and, taking hold of him with her mouth, scratched his soft skin, leaving a single spot of bright red blood.

As before Calgacus stood still, waiting, and watching his sister. I could see no sign that he was willing to give up his game. With a deep sigh Susan levered herself up out of the soft depths of her chair, and went to Calgacus, picking him up, briefly hugging him as she did so (which made him struggle to be free), and placed him in the kitchen, away from the other puppies until he was able to calm down. Nothing was said, but I got the impression that this wasn't the first time the little scenario had played out.

In those early meetings, before Calgacus had even come home, he was teaching me about himself. His dislike of being picked up and hugged, and refusal to sit on knees or have close contact made me realise he had likes and dislikes. He was young, but even so he wanted autonomy; to be listened to. I wanted to hug him close, but it wasn't all about me – something he made clear from the outset. Similarly, he demonstrated that he would not be pushed around or told what to do – even by a much-loved authority figure like Lola. He didn't recognise leadership based on strength.

Every time I met him, the desire to share my life with him grew. I could see and feel what a special puppy he was, and soon, he was all that I could think about ...

# 5 Early days

I'd read obsessively about dogs but none of it prepared me for the reality of living with a puppy.

I was shocked that Calgacus' needs were so demanding; the reality of that responsibility weighed heavily. I look back on that time in my life and regard it as a mixture of joy and misery. I remember surges of overwhelming love for Calgacus. Smiling, I'd watch him sleeping, and feel my chest fill with lightness and a desire to know more about the chubby puppy resting beside me twitching, as he dreamed about whatever wonders had caught his attention.

There were other – very different – times. When awake, Calgacus would chew furniture, clothes, shoes – anything within reach. He would steal food if it was lying around. Getting into the habit of tidying everything out of the way was exhausting: I'd never had to pay so much attention to my surroundings.

He needed to be taken outside to the garden at least once an hour – more often if he'd been playing, had just woken up or had eaten – to go to the toilet. I spent vast amounts of time worrying about his toileting, trying to help Calgacus get it right, and cursing myself every time it wasn't. Home life revolved around Calgacus' bladder and bowels, and every successful trip to the garden was celebrated as if something wondrous.

Other times there was no cause for celebration, when all Calgacus felt like doing was chasing leaves in the garden. I'd be fooled into thinking he didn't need to go and take him back indoors, only to watch a bright yellow puddle of urine spread across the floor as soon as he was inside.

He needed two or three short outings a day to stretch his legs, and practice the things I was trying to teach him, and time in which to meet people, other dogs, and see as much as possible of the world. These excursions, rather than taking place when I fancied a walk, had to occur throughout the day. Before Calgacus, I was in the habit of walking every day so hadn't anticipated this being a problem, but I found that there is a big difference between once-daily walks that I could skip if I wanted, and compulsory multiple walks every day.

Until Calgacus I'd had a life that demanded little of me: other than turning up for work, I was pretty much free to do whatever I wanted. Back then, I would regularly have days to myself in which to sleep, read, eat, and watch TV – not leaving the house or doing much of anything other than relax. It sounds like an indulgent way to spend my time, and it was, but I enjoyed those days, looked forward to them, and tried to fit them in whenever I could.

I often enjoyed time to myself outside the house, too, spending hours sitting quietly in coffee shops and bars with a book. My partner would regularly

# Canine aggression

ask me how I managed to use up half a day doing the shopping. The answer, of course, was that I'd get the shopping and then drift for hours, wending my way from coffee shop to coffee shop, reading and thinking, before returning home. I could pass a Sunday afternoon with ease by going for a walk, and then rewarding myself with a large glass of red wine; sipping and reading.

This all had to stop when Calgacus came along, and, instead, my days revolved around the next walk, and whether he had recently been outside to go to the toilet. Would the next trip involve a frustrating wait followed by an unpleasant mess to clean up if I brought him back inside too soon?

The adjustment was hard. This was something I'd chosen, for sure; something I'd planned for and dreamed about; something I'd wanted my entire life, and yet, frequently, it made me feel depressed. I vividly remember one particular walk with Calgacus in a field, where, imagining the future, I saw the next decade or so of my life, and didn't like what I saw. On that sunny autumn day, watching my russet-and-black puppy chase flame-orange and poppy-red leaves, I slumped, wanting just to go home, get into bed, and stay there. Only I couldn't because toilet trips and more walks filled my day, and then the next day and the one after that: weeks, months, years of this every day.

I looked at Calgacus and, instead of seeing the gorgeous puppy I felt so much love for, I saw a burden; an obligation I was no longer sure I wanted to fulfil. I cried on that walk, letting the tears come and getting them out while it was just Calgacus and I, and I wouldn't have to explain myself to anybody else. And there was guilt about my feelings, too. Neil and Susan had let me take home their favourite puppy, and here I was feeling down about having to care for him.

I'd like to say that those feelings disappeared quickly but – I won't lie – they didn't, and, over the course of the following year I felt this way a lot. What I didn't know then – and which might have made all the difference – was that I wasn't alone or unusual in feeling this way. Subsequently, I've spoken to others who describe feeling the same way when they took on a new dog. Even those who've had lots of dogs have spoken of  feelings of anxiety and worry when a new one joins them. I wish that some of the books I'd read about dogs had mentioned the possibility for feelings of depression and anxiety after bringing home a puppy.

Many of the things that Calgacus did – normal puppy things – were sometimes challenging, and chewing was one. A normal activity for dogs and puppies to engage in, chewing is necessary for their physical and mental health: keeping teeth clean, exercising jaw muscles, and providing relaxation. Additionally, puppies learn about the world and the objects in it by using their mouths in this way. Chewing when teething helps teeth to come through, and also alleviates the pain associated with it.

The books I'd read concentrated on how to train dogs to fit into a human world, and had little to say about what a dog – and especially a puppy – needs, but Calgacus taught me.

He stripped the wallpaper from much of the living room up to around waist height over the course of a few weeks, then moved on to experiment with the sofas, joyfully chewing through the cushion covers to discover mounds of squashy, chewy filling. Tail wagging he dug into this with his feet, and then reached down to grab a mouthful, pulling it out and prancing around with it, tossing it in the air and pouncing on it like a cat. Each time I stopped him in his tracks, although I know now that this isn't the right thing to do, as Calgacus responded by simply waiting to chew until he was alone at home.

Wood was fun stuff to chew as well, and he was once found lying on top of the dining table, contentedly chewing on the window sill beside him. He never tired of chewing wood, and even when he was eleven years old, if he should knock over a small table over while playing, he would chew on the legs.

And neither were my books safe – especially the hardbacks – and, eventually, all were moved into the spare bedroom, from which Calgacus was banned.

I know that, with some dogs, destroying things can be caused by separation anxiety when they are left alone, but I don't believe that this was the case with Calgacus. Rather, I wasn't doing a good enough job of giving him things to chew on. What I should have done was give him an endless supply of different sorts of toys and chews to let him explore, tear apart, play with, exercise his jaws, soothe the aches from his teeth growing in, and enable him to relax. Once I realised this and began giving him a variety of toys – including plenty with food inside – that he could chew when I went out, the furniture destruction stopped ... but not before I'd had to replace two sofas and four armchairs.

I suspect that if I had had a little more canine knowledge, I could have encouraged Calgacus to chew more appropriate things, but I'll never know. He and I were learning together, and I didn't then have the knowledge that I have now.

In the house, chewing and house training were the main things to deal with, and on walks, it was meeting other dogs. Calgacus had been overly-forceful and bullying toward other dogs when he was still with his litter mates, and that attitude continued once he came to live with me. He liked other dogs and wanted to be friends, but if he wanted to play, he pretty much didn't care what the other dog wanted. For many puppies, learning to be polite occurs through their interactions with litter mates and other dogs, who may refuse to join in games by walking away from or ignoring the puppy. A show of teeth or barking may also be used if the puppy is particularly pushy. Most puppies will back off when confronted by an irritated adult dog.

Not Calgacus, though, who had found it easy to ignore his mum's attempts to get him to tone things down, and most dogs we came across didn't have the size and strength that Lola had. On walks he'd skip away from irritated dogs, bounding happily, pretending that they were joining in with his game by chasing him. Dogs who cowered or lay down when he approached would be pawed or nudged in an attempt to get them to run.

# Canine aggression

Even I could see that the only one enjoying these 'games' was Calgacus. Suspecting that this issue would get worse as he grew larger. I tried a new tactic on our next walk.

We met a blonde, fluffy dog, and Calgacus ran up to her, tail wagging. He barged her with his shoulder, turning as her teeth flashed at him, and running off. He looked back as he ran, checking if she was behind him, and when he saw that she was not, he looped round, running back to her, angling toward her shoulder for a second barge.

"Calgacus," I called, stepping toward him.

I held the lead in my hand, and my intention to attach it to him was clear. He dodged, jumping aside at the last moment to avoid me.

I plodded through the muddy field after him, moving much more slowly than a lively puppy.

"He's a bit rough. Sorry," I said to the other dog's carer, who was watching with an amused grin on their face.

Eventually, Calgacus stopped to sniff something and I caught up to him. I patted him and handed him a piece of sausage when I clipped on the lead, and we walked briskly away from the blonde dog.

I kept up this new tactic while also looking for dogs that did enjoy playing with him. Living near us were a little Collie cross and a Bearded Collie, both of whom were happy to play with Calgacus. When he was very young, he had a friendship with an amazing Neapolitan Mastiff – a dog so much bigger that he barely noticed if Calgacus ran into him or grabbed hold of his jowls.

Relationships with other dogs were not the only source of stress when it came to walks. There were also deer. I have friends who tell me about the joy of seeing deer when they walk, their ethereal beauty seems to catch the imagination: slender faces that are all-eyes and curiosity, gone in a blur of reddish fur. They make for a magical sight, transforming a boring day into one filled with wonder.

For people with dogs, deer can be tricky, however. They are incredibly exciting to lots of dogs, many of whom cannot resist the temptation to chase a running deer. But dogs chasing deer can result in trouble with the law, and cause enormous anxiety for their carers, should they pursue too far and become lost. For the dogs, there is the risk of injury, becoming lost, or developing heat exhaustion, and the deer suffer, too, from the fear and stress of being chased. Deer-chasing is a particular issue in Scotland where, it seems, there are few places without them. I've seen them in city parks, industrial estates, and in tiny green areas in the middle of housing estates.

In my experience avoiding them is impossible, but I hadn't realised this until Calgacus was approaching his first birthday, when he developed a fascination with them, leaping fences and running across fields to chase the animals.

Sometimes, he found it hard to get back to me, and often I was scared by how far he would run. I got into the habit of walking in places away from roads and livestock, to try and keep him as safe as possible if he did chase a deer.

One morning he and I were walking in some beautiful woodland close to my home. The narrow track we trod was bounded on one side by a steep drop to a river, and on the other by dark, tangled forest. The sun had just come up, revealing a blue sky filled with wispy, pink-tinged clouds. I breathed in the clean, pine-scented air and felt a moment of peace. Looking up, the moment was shattered by the sight of three large deer on the track ahead. For a moment, everything stopped and none of us breathed, then the deer turned and ran, their white tails bobbing through the forest, with Calgacus, a smaller, striped figure, running behind, his strides quickly outpaced by those of the deer, as they disappeared into the trees. I stood alone on the path and waited. I expected it to be a short wait. Calgacus ran much more slowly than any deer, and, although he found them exciting, his attention would falter once he could see them getting further away.

Behind me I heard thuds and rustles and branches breaking with short, sharp cracks, and turned to see two deer run out of the woods straight at me. Behind them by quite a way ran Calgacus, his head up and his eyes bright as he galloped along. The deer were immense. Seen at such close quarters they lost their otherworldly, slender appearance, and became large, powerful animals, their muscular shoulders bunched and tensed as they ran, their breath puffing out, clouding in the early morning air. I tensed and stared at them, then looked down at the long drop to the river, wondering if I would survive being knocked down the drop by a pair of panicked deer.

That nightmarish moment stretched out as fear distorted my perception of time. At the last moment the deer turned, plunging down the slope toward the river. I watched them run, sure-footed, never stumbling or tripping, down the slope I'd been sure would be the scene of a serious accident for me.

Calgacus stopped when he got to me, breathing hard, tongue out, looking for some food. I bent down to hug him, close to tears with relief.

Other challenges to my fumbling attempts to help Calgacus grow into a well-behaved adult dog came from people. One sun-soaked weekend day, I walked with Calgacus along a path that opened out onto a wide field. Sitting on the grass, in the middle of the field, was a gang of young men dressed in dark green or camouflage clothing, drinking from cans of beer and bottles of that Scottish favourite, Buckfast. Smoke from the cigarettes they held drifted above their heads as they laughed with each other. Beside them lay air rifles, and some dead rabbits.

Calgacus was interested and a bit startled to see them. He loved making new friends, and finding a bunch of young men lying on the ground was a new way to meet them, so he went straight over to say hi. One of the men smiled, reached across the grass and picked up a rabbit, which he held out to Calgacus and smiled at me.

# Canine aggression

"He'll love this. Fresh rabbit. Dogs love it."

Calgacus took the rabbit, went into a bush and ate as much of it as he could. From then on, finding dead animals to eat was a favourite pastime for him, and a constant source of annoyance for me. A positive slant was that this encounter gave Calgacus an even greater love of meeting strangers. Happy social interactions are something that dogs need – especially those who may otherwise become suspicious of strangers.

With the benefit of hindsight, I can see how rich an educational experience my first time with a dog was, and was able to experience first-hand just how hard learning new things can be, both emotionally and psychologically. Had I appreciated this in advance, I think I would have been kinder to myself, and felt less anxious about the ups and downs I experienced. This was a time of learning and development for Calgacus and I, which helped us to bond.

Before finishing this chapter, I have one more story about a challenge from those early days.

Out of all the adjustments to my lifestyle and the struggles I experienced, one of the hardest was the dark, of which I am terrified. Working full-time as I did, morning and evening walks had to be done in the dark, and I had to force myself to do this. One winter evening, we set off to walk around the field at the back of the house. I took a heavy plastic torch with me, and clutched it as we walked. The bushes – menacing shapes – drew close, their branches scratching me and grabbing at my hair. My breath fogged in front of me in the light of the torch. I shivered, feeling the dark press around me. Calgacus was having fun, running around, sniffing at scent trails, obviously with no concern for the state I was in.

The torchlight caught something that wasn't leaves – something orange tucked into the bush at head height. I took in more as the torch moved. A hand. A gun in the hand. The orange something resolved itself into a balaclava covering the head and face of the man holding the gun.

The darkness grew more oppressive and, for a moment, I thought I would scream. My hands trembled. What should I do? Club him with the torch? Run? Where was Calgacus? The dog I'd been assured would become protective of me.

In the end, I fell back on old habits and spoke. "You gave me a real fright," I said and laughed – the sound thin even to my own ears.

"Oh, sorry!" He took off the balaclava to reveal one of the rabbit hunters we'd met that sunny afternoon. "I came out to hunt. I didn't think anybody would be out here now. When I heard you coming, I knew anybody who met me would get a fright so I thought I'd hide in the bushes until you went by."

I laughed, took a deep breath, aware of the night-time scent of the plants, complex, earthy, and wild. "No worries. I'm just glad you weren't a serial killer waiting for your next victim."

A rustling, cracking sound from the river bank signalled where Calgacus was, and he ran up the bank to see what was going on. I wondered if he might stand between me and the man with the gun or do something else that might indicate a desire to protect. He did neither, not even glancing at me, but greeting the man with enthusiasm. The provider of rabbit meat had no food to share this time, so, instead, he made a fuss of Calgacus, scratching his ears and talking to him, telling him what a big, fine dog he was.

The memory of that walk scared me for months. How I wish that this was the most frightening thing I experienced with Calgacus.

Visit Hubble and Hattie on the web: www.hubbleandhattie.com
hubbleandhattie.blogspot.co.uk
Details of all books • Special offers • Newsletter • New book news

# 6 Getting better

Attending classes can be an excellent way of learning more about how to teach dogs, in a supportive atmosphere, with other people who are also learning, and, once Calgacus and I got into the swing of our first class, this became an important part of our lives. I learned how to communicate better with Calgacus, made some friends, and started to feel as if things were coming together. I still consider myself lucky to have discovered that class when I did – even with the problems it was to cause us.

I can remember the fluttering in my stomach, and how my hands shook at our very first class, or, more accurately, the assessment with the club's head instructor that was designed to determine which class would be best for Calgacus and I. I wanted the instructor to like me and my dog. I knew that Bull Mastiffs are viewed negatively by many, and I was keen to show that my dog was a good dog, and that we could come to class.

The head instructor was called Ursula: a pretty woman with long, blonde hair, and skin that glowed with health. Her knowledge was apparent as she spoke to me, and she had that rare ability to explain complex concepts in an easy to understand way that helped put me at ease. During that first session, she showed me how I might go about teaching Calgacus, and demonstrated this by getting him to walk to heel, and lie down when asked to. Her movements and speech were sure and confident, and I found myself nodding as she spoke, gaining new understanding.

Calgacus, too, seemed to pay attention, watching Ursula to figure out what she wanted him to do (I'm sure she was the reason for his life-long love of blonde-haired lady dog enthusiasts!). At one point in our session, Ursula described Calgacus as an old soul, telling me that he seemed to already know some of what she was teaching, which, of course, he didn't.

I knew little about the dog world, and nothing at all about the wide variety of sports and activities that people take part in with their dogs. It would never have occurred to me that people would want their dogs to walk so precisely next to them, or sit close in front of them and then run to their left-hand side, for example, when asked.

By the end of the session, it felt like Calgacus and I had passed a test, which meant that we would be allowed into a class where we could learn more. I left that day excited and pleased with how well Calgacus had done, and looking forward to starting class.

A few weeks into classes, I realised that something had changed for me: I was finding our walks more and more enjoyable. Sometimes, I would be surprised by a welling up of happiness starting in my stomach and bursting from the top of my head: a feeling so powerful that I would stop walking to concentrate on it.

Morning walks were our most special times, wandering in peaceful woods, watching the sun dye trees rich shades of orange and pink. Calgacus would run around, young and full of life, charging between the trees, delighted if we met another dog he could play with, and, if not, just happy to be outside. We'd practice some of the things he was learning – coming back when called; staying still when asked – anything that I felt was useful for Calgacus to know how to do, or which we'd been practising in class.

I'd walk on in that state of euphoria, falling in love with the world, with Calgacus, and with my life. This was something special; something important.

I'd felt that way before, of course. Falling in love with a boyfriend was a similar feeling. Getting a job I really wanted and hadn't expected to get felt a bit like it. Passing difficult exams. Those first few runs at the start of a new ski season when I was a teenager.

The difference here was the strength of feeling. I'd never felt that euphoria so strongly before. Nor had I felt it so reliably over such a long period of time. I was used to happy feelings being fleeting in my life, drowned by anxiety after a short time. This time the euphoria went on and on.

Even Calgacus' tendency to overwhelm and bully other dogs was diminishing, as he gradually learned that if he upset another dog, I took him away from that dog. Over time, he learned to be careful and moderate his playfulness so that other dogs were not frightened of or annoyed by him. Eventually, he was gentle enough to go on walks with my neighbour's elderly dog, and sufficiently socially skilled to walk with another neighbour's well-mannered dog without annoying her. He was happy to have so many friends, and I was equally thrilled that he could play with other dogs as much as he did.

Another positive change was that the more he and I practised the training he'd received in class, the less he stopped running off to try and meet new dogs; being happy to stay with me and practise – in exchange for sausages – if we met a dog who didn't want to play.

Things were looking up, and I was feeling pleased and proud of myself for having helped Calgacus learn to be more gentle. Had I known what was coming, I'd have more firmly held on to those moments, and appreciated them more.

Our classes became a highlight of the week: evenings filled with friendly people and fascinating learning. On class night, I would whisper to Calgacus that it was time for his special night. One night the dogs learned how to skateboard, which involved many pieces of hot dog, lots of slobber, and much laughter. Over the course of that class, Calgacus learned that he had to put one front paw on the skateboard with sufficient force for it to move. At the end of the first year, we received an award for being the most improved in the time we'd been there.

I developed a hunger to learn more about dog training. Ursula and some

# Canine aggression

of the other instructors talked about conferences they'd been to, people they'd met, and things they had learned, and I wanted to be part of that world. The club ran many classes – for puppies; dogs who were improving, an advanced class (which Calgacus eventually attended); a special one for retired Greyhounds, and a class for dogs with behavioural problems. In time, I was invited to begin learning to be a class instructor, which I eagerly took up.

I still had little knowledge of dogs, only the basics of how to teach, really, and was quite frightened of those who showed any kind of aggressive behaviour. I can remember watching Ursula doing a one-to-one session with a young Labrador, who was there because she barked at strangers. Anxiety bubbled up in my stomach, turning it into a churning pit of fear, and I was unable to go into the same room as the dog, even though she was on a lead.

For Calgacus and I the classes were all fun and sausages, but this was not the case for dogs like the young Labrador, who had a quite different experience. There were still sausages for getting things right, but there was also shouting, water sprays and yanked leads. Ursula explained to me that the best way to teach a dog not to be aggressive is to make it clear to them that if they are, something awful will happen to them, so that the dog will be too terrified of the consequences of showing aggression. There is a certain bad logic to this kind of thinking, and it made sense to me at the time. I carried on taking Calgacus to class.

However, I didn't think that this way of training looked at all nice. I watched as people were taught how to shout at their dogs loudly enough to scare them, and knew that I couldn't do that. It seemed as though this sort of approach was necessary for some dogs, and I felt badly about not being able to help those dogs. I hugged Calgacus sometimes, feeling glad that I didn't have to use this method with him as he didn't have those sorts of problems.

# 7 So, what *is* clicker training?

Despite my reservations about some of Ursula's methods, Calgacus continued to benefit from the classes.

On that very first meeting, Ursula showed me how to use a clicker as a teaching tool. The clicker method is based loosely on the work of B F Skinner, one of the most influential of American psychologists. A behaviourist, Skinner developed the theory of operant conditioning, based on the simple premise that living beings are more likely to repeat actions that have consequences they find pleasurable, and less likely to repeat those that have consequences they find unpleasant.

Of course, like all simple concepts, this one quickly becomes complicated with a little thought. Consequences can be tied together in ways that are hard to understand – even for the individual experiencing them. Drinking and eating habits among people are a common example of this. I love nothing more than meeting friends socially, to share food and a few glasses of wine, and derive an enormous amount of pleasure from spending time in that way. That's the good consequence. The not-so-good consequence is that too much wine makes for a hangover the next day, and too much food makes for weight gain. The consequences – good and bad – are intermingled, and the balance between them is complicated. For me, the pleasurable consequences outweigh the times I overdo it and get a hangover, as well as a tendency to be overweight. Other people might decide differently when faced with the same choices.

To further complicate this, emotions matter, and Skinner's work has been heavily criticised for failing to take this into account. By focusing on what people and animals *do* rather than how they *feel*, Skinner has been accused of attempting to reduce people and animals to emotionless machines at the mercy of the environment they operate in.

I have friends who shudder when I've explained to them what the clicker method is.

"But that's behaviourism," they say, sometimes glancing around self-consciously, as if I'd suddenly farted and they're worried in case somebody thinks it was them.

If I viewed Skinner's work in similar terms, I suspect I'd feel the same way, but I don't, and lots of people who work with animals that I respect and look up to don't, either, and instead, have taken Skinner's work and built on it. Rather than regarding dogs as emotionless machines that can be manipulated into doing what is required, clicker enthusiasts see dogs as sentient, emotional, and self-aware beings who need training that is appropriate for them as individuals.

# Canine aggression

The clicker itself is nothing more than a small, plastic device that makes a distinctive clicking sound when pressed. The usual way to give this meaning is to make sure that the sound indicates good things will happen. When dogs are first learning what the sound means, a special treat – soft, fragrant food such as sausage, roast meat or cheese – is given immediately after the click, and thus, the dog learns that the sound is followed by something very nice. What this is differs from dog to dog, of course: some particularly like cheese, others may prefer roast lamb, or, if food is not their motivator, a particular kind of touch, or a game with a toy. Spending time working out what each dog likes is a vital part of this teaching method. Once the dog understands the link, the clicker can be used whenever the dog exhibits the desired behaviour.

It's vital that the click occurs at the exact moment that the dog does what is being asked of him. So, if jumping over an object is what is being taught, the click should occur when the dog is in flight, almost as if taking a photo of this event. This simple device, and the related thinking that builds on Skinner's work, has opened the door to some radical teaching and learning when it comes to dogs, and, in particular, an appreciation and focus on the understanding that learning anything new is an emotional process. Frustration is the emotion I hear mentioned again and again in relation to teaching dogs. A dog who isn't sure what he is supposed to do can easily become frustrated if he doesn't hear the click very often, which can result in him being too stressed to learn effectively. This is such a strongly recognised issue now that articles, blog posts, and YouTube videos have been produced, and class training instigated, on the subject of how to recognise when a dog is becoming frustrated, and how to prevent this.

And, of course, it's not just the dog who becomes frustrated. I see it in people, too; have even experienced it myself when learning something new. Frustrated human students are no more able to learn than frustrated dog students, and can become disruptive, or simply shut down and give up. And, when that happens, and if everybody else is coping, the conclusion is that there must be something wrong with them.

Those who teach dogs using clickers don't see things that way, happily. When a frustrated dog barks or grabs the teacher's clothing, or refuses to participate, they do not consider that the dog is at fault, but, rather, themselves. Even if they've previously taught many dogs using the same methods, and never had that response, they don't blame the dog. Spotting the small, early signs of frustration – little whines and hesitations; whacks with paws – and adjusting their teaching accordingly, is the course they take, appreciating that understanding the impact that emotion has on teaching is a central part of what they do.

There is the question of free will, as well. Skinner's work is criticised for denying free will by reducing living creatures to biological 'machines.' Whether or not that is how Skinner understood things, it is not the view of many of those who make teaching non-humans part of their lives. Those of us who love them understand that dogs are individuals with their own likes, dislikes, fears, and joys. The question about free will has become not so much about whether or not dogs have this (they do), but more about how to let them express and evolve it more fully, and how encouraging them to exert free will and explore the world in their own way can be helpful in a teaching context.

A couple of issues make this difficult. For a start, the balance of power in the relationship is heavily weighted toward us. We decide what our dogs eat and when. Where they exercise, for how long, and in what way. We decide what our dogs are taught, and how they are taught it. This being the case, it is certainly feasible that dogs have lives with very limited or no free will at all, and many do, sadly.

Moreover, it's not very easy, sometimes, to allow our dogs increased autonomy. In the UK it is regrettably common for dogs to be euthanised for communicating in ways that are normal for them – growling, barking, and even inhibited biting are all normal facets of dog communication, but they make us so uncomfortable that dogs are sometimes thought to be too dangerous to be allowed to live.

I can remember once meeting a dog who was living in a rescue centre. She was a pretty little brown and black mixed breed with elegant, long legs and dainty feet. I observed her walking around with a person she didn't know, staying close to them when they walked, and sitting next to them when they stopped. If she was barked at by another dog, she politely turned her head away or hung back behind the person with her. She avoided crowding the other dogs as much as she could, moving to the other side of the person when approaching them.

After a while I asked a member of staff why she was there. She looked so sweet and was behaving so well – who could possibly have given her up? The answer was hard to believe. Her previous family had taken her to the vet to be put down because she'd barked at another dog while being walked directly toward the dog on a narrow path. She had barked. That was all. Her family had been so scared by this behaviour that they felt she was too dangerous to live. Luckily for this gentle little dog, they were talked into giving her to the rescue centre, where she had the chance of finding a new family.

Dogs live emotionally rich lives. They love; they care for their offspring, and form close bonds with us, other dogs, and, sometimes, other species. If we choose to have them in our lives, I believe we owe it to them to think about their emotional well-being, and find ways to give them as much freedom as possible to live their lives as they choose.

The people who are my inspirations and guides in this respect use their clickers to help dogs learn things that will keep them safe, and do so in ways that encourage the dogs to explore while learning; trying out different things and expanding the boundaries. With a skilled teacher and a dog who's had enough time to explore, the results can be amazing. Dogs have learned to guide the blind, sniff out drugs and explosives, find a lost person on a snow-covered

# Canine aggression

mountainside, provide support to people with physical limitations, and bring comfort to the mentally ill.

Good teachers will use their observations of the dog's learning to work out what that dog loves doing the most, and try to find ways of encouraging and developing those interests. This is powerful stuff, and, excitingly, can be applied to humans, too. I watched an interview recently where Kristine Barnett, founder of Jacob's Place, a non-profit organization designed to help children with autism, as well as an award-winning sports league for autistic children, spoke of how she taught by concentrating on allowing her son, Jacob, to explore the things he was interested in. Kristine had been told that Jacob's autism was so severe he was unlikely to ever learn to speak, and would probably be dependent on her for life. Kristine's teaching has allowed him to develop an interest in physics, on which he gives talks: according to media reporting, Jacob is on-track for a Nobel prize nomination.

Being taught in a way that encourages exploration, and the following of interests, is much more likely to result in students who have a genuine enthusiasm for learning, and engagement with the things they are exploring. In their book *We Make the Road by Walking* (http://www.temple.edu/tempress/titles/804_reg.html), Myles Horton and Paulo Freire – two of the most prominent thinkers on social change in the twentieth century – speak about this, and it's clear that both of them wanted to make education and learning so attractive that people wanted to engage with it. They wanted their students to feel empowered by their learning, and, in so doing, they created educational practices that I see echoed in the teaching practices of many of the people I respect in relation to their attitudes toward teaching dogs.

Of course, exploration of and encouragement to develop free will has a flip side. Self-aware, self-confident, exploratory beings are not always obedient. They question; they sometimes say 'no' to things, and aren't always easy, biddable, non-critical companions. But, to my mind, this is a good thing, because I want our lives together to be long years of exploration and discovery about each other. I want to know my dogs just as well and as lovingly, as I do the people in my life.

# 8 All change

Things were going so well with Calgacus that I wanted to get a second dog: a friend for him: another dog to love and watch grow up.

Enter Katie: a soft, strawberry blonde who had a cute snuffle when she was excited, and came from an experienced breeder of Bull Mastiffs.

We drove one afternoon to pick her up and, when we got home, Calgacus could not have been happier. A puppy! All of his own! I don't think I'd ever seen him so happy. With us watching nervously, he got to know his new companion, who was only about the size of his head, so we worried that he might accidentally hurt her.

Katie's arrival brought with it a need to learn much more about how dogs communicate with each other. Having two dogs in the house is not at all the same thing as having one dog. This period in my life taught me another lesson – things change; no matter how much you might wish they didn't, they do, and often in unforeseen ways.

Almost immediately, the commonly held expectation that dogs will always behave in the same way was shown for the fallacy it is. I don't know any people who aren't at times joyful, exuberant, playful, sad, angry, frustrated, so why should we expect dogs to always show (or not) the same emotions? Our dogs need to behave well in society, otherwise there's a risk they will be labelled 'vicious.' Yet, being the sentient creatures they are, they experience a range of emotions, just as we do, and display different behaviour as a result, just as we do.

So, dogs can bark and growl without being considered 'bad.' They are complex characters, who can behave differently depending on how they feel or what is going on around them – just like we do. This, of course, adds a level of complexity to choosing to live with dogs. It isn't enough to look for a 'good' dog and then go about life without considering the dog, as he actually needs his people to make space and time to learn to understand him.

I'm reminded of a dog that Calgacus was friendly with – a large, happy, bouncy chap. He and Calgacus enjoyed many a game together, and were always pleased to see each other.

One morning I was walking with Calgacus, and we met this particular dog. Close to him was a smaller dog who I had never seen before – a little Collie cross with dark brown fur and a white patch on her chest. That morning, instead of wagging his tail and moving toward Calgacus with a soft, playful expression on his face, our friend ran toward Calgacus, his tail held high and rigid over his back, barking loudly. He stopped about twenty steps away, and growled low in his chest. It was clear that he didn't want Calgacus going close that day, and so I

# Canine aggression

quietly clipped on his lead, and we stood still for a time. The dog's carer appeared on the path, and he turned and went to her.

"Was one of them barking?" She asked.

"Yes. I'm not sure why but he really didn't want Calgacus near him." I told her.

"Ah – it might be because of Milly here," she pointed at the new dog. "She is my daughter's dog and they are visiting just now. He can get a little bit protective of her. Sorry if he scared you."

Suddenly, the reason for the change in this dog was clear. Everything was different for him that morning. There were visitors staying in his home – which I have no doubt he enjoyed but which also involved a change to usual routines and habits. He may have been tired or more excited than usual, and he also seemed to feel either a responsibility for Milly, or didn't want to share her attention. Whatever the reason, this dog was not comfortable with Calgacus, and let us know this in an appropriate and safe way that Calgacus and I took note of, respecting his wishes. The next time they saw each other, they had their usual game together.

That incident was a powerful example of why we should not always assume that dogs will behave in the same way, regardless of what is going on in their lives.

This story was a digression to the account of Katie moving in, but is relevant to what came next. Calgacus' excitement at having Katie move in continued until dinner time, but then his mood became dark. Katie had a crate in which she had her meals; somewhere safe for her to eat in peace, and in which to rest when Calgacus' enthusiasm for playing became exhausting.

I placed his large bowl in front of Calgacus, and he dived in quickly, jaws working and eyes focused on his bowl. Then I walked to the other end of the 21 feet long living room to put Katie's bowl in her crate.

"Here you go, sweetie," I said, watching as she stepped forward to eat.

Behind me I heard scrabbling on the floor and a low, guttural growl. I turned. Calgacus was racing across the room – moving so quickly that his paws could not get purchase on the floor, and were scrabbling, toes clenched. His brow was wrinkled and his lips pulled tight, showing some of his teeth. He opened his mouth and barked, showing all of them.

I froze – shock making my limbs heavy. Unable to move, I watched as Calgacus completed his charge across the room and ran headlong into Katie's crate. When I recovered, I got him out of her cage, and moved to stand between him and the cage, pushing him aside, then sending him back to where his own food lay untouched. I stood there, trembling, mind racing, feeling sick and sad, and utterly unprepared for the realities of caring for more than one dog.

My shock was such that I have memory blanks about this incident. One of my friends who read an early version of this book asked: "What about Katie? How did she react to this? You don't mention her at all ..." and I read back to find that this is true. I simply cannot remember what Katie did or how she looked. What I can remember is that Katie finished her meal and Calgacus finished his – wearing his lead and with me holding the end so that Katie was protected from his wrath. I turned over the incident in my mind, worrying at it, unable to rest that night as I struggled to understand. In those dark hours, huddled in the warmth of a thick duvet, I found a question. What if Calgacus just didn't understand that there was enough food to go around? What if, in his mind, this was something of a life-and-death struggle?

I had nothing else to go on so decided I would try and figure out whether that might be the case. At the next feeding time, I was prepared. I attached Calgacus' lead to his collar, and tied the other end to the door handle near to where he was fed. I gave him his large bowl of food, then walked across the room to feed Katie. This time, Calgacus looked, growled, and pulled toward Katie, but couldn't charge because he was restrained. Moving quickly, I went and stood next to him.

"On you go, sweet pea, eat," I told him, opening the packet of ham I'd taken from the fridge and dropping a piece into his bowl.

He lowered his head and ate, gulping at his food, sometimes glancing toward Katie. He finished his meal first, and I stayed with him, dropping pieces of ham into his bowl, keeping up a constant supply until Katie had finished her meal. I left Calgacus tethered while I cleared away the bowls.

The next day I did the same. And the next. Calgacus grew calmer, and stopped growling and glaring at Katie, eating his own food, and then looking to me for ham.

On the third day, I repeated the exercise but without tethering Calgacus. By the end of the week, he was calm and settled while Katie ate, showing no signs of concern. Soon after that, I was able to stop standing next to him. Sometimes he would come to me for extra food, which I would give him, but eventually even that ceased.

Things were different for Calgacus in our home with Katie living there, too, and even when the mealtime problem was resolved, there was still work to be done in getting used to a new dog. From the beginning, Katie was not like Calgacus. My poor Katie. Her life was plagued by health problems and anxiety, and I was reminded while going through old emails when writing this book that she battled those demons even as a puppy.

On my first walk with Katie, we met another dog. I smiled and waited for Katie to approach the other dog for a game. Instead, she ran – a chubby, gingerish blob of fast moving legs and tucked tail – into a bush, where she stayed until the other dog was gone. On another of those early days, I took Katie outside our house on a lead to meet another puppy who'd moved into the street. She couldn't run away that time, but, instead, barked and growled to keep the other puppy away.

I'd left Calgacus at home for those early walks, wanting to spend the time

# Canine aggression

bonding with Katie, and give her the time to form friendships of her own. When that didn't go as well as I'd hoped, I wondered what might happen if Calgacus was with us, so I took him on one of our walks. We met dogs we knew, and he greeted them with joy, running through the grass, scattering the young, spring plants as he and his friends chased each other around. We repeated this a few times a week, and, gradually, Katie learned that other dogs could be fun. Tentatively at first, she began to play, much slower and more carefully than I had ever seen Calgacus behave – but she did play.

I can remember the relief I felt watching her first few games, and the joy I experienced the first time she played chasing games for so long with a pretty Collie cross that she had to lie down and rest. That afternoon, lying on bright green grass, her tongue hanging out and her body relaxed, she seemed so happy. Katie never developed the same level of confidence that Calgacus had, but his presence on some of her early walks was an enormous help.

Calgacus was about 18 months old when Katie moved in. As Calgacus wasn't neutered, Ursula encouraged me to take him to get the operation done. She thought it would make life easier for both of us, and I was inclined to agree with her.

The timing of neutering of male dogs is a subject in itself, debated regularly, and with much written about it. The reason I was inclined to agree with Ursula when it came to Calgacus is because, so often, entire male dogs – especially adolescents – run into trouble with adult dogs. Calgacus was no exception, and I was upset to find that he reached a point – at roughly a year old – when his presence alone seemed to irritate some dogs, who would approach and then barge him or jump on top of him, in an effort to intimidate. This didn't happen all the time by any means, but there were probably ten or so incidents in about six months.

Calgacus didn't seem upset by these, however. Quite the opposite, in fact, as he seemed to regard these mildly hostile interactions as an opportunity to be rougher with other dogs than I'd taught him to be: pushing, shoving and arguing to get what he wanted. Certainly, he bore no hostility to the other dogs afterwards; there were never any injuries, and it seemed fairly good-natured all round. He didn't go around picking fights that I could see – or maybe he did and I didn't have the skill then to notice. I wasn't overly worried but, all things being equal, I probably would have had him neutered then but for Katie.

Having her in the house required adjustment for everyone, and she'd needed some help and support to deal with the outside world. I wanted to be sure everybody was settled before putting Calgacus through even a minor operation, so we waited. I expected to be able to take Calgacus to be neutered a few weeks from then.

As it turned out, this was too late, and the events one night in training class signalled the start of everything falling apart.

Calgacus was lying quietly by my side, on his lead. The instructor stood a distance away, giving advice to somebody about how to get her dog to do an exercise. While everyone was caught up in their conversation, the dog – a large, adult Labrador off his lead – came over to Calgacus and stood stiffly, tail up, and looking down into Calgacus' eyes. Back then I didn't know enough about dogs to understand what this means, but I do now. This was the canine equivalent of one person saying to another: 'Come on, then, if you think you can.'

I could feel the change in Calgacus, who tensed, his muscles becoming rigid under his velvety coat. When the dog's carer noticed what was happening, she called to her dog, repeating her call several times in the space of a few seconds, but her dog wouldn't move away from Calgacus. I was sitting in a plastic chair pushed against a wall, and couldn't see any clear escape routes from this situation. Finally, the dog's carer came over to get him, reaching for his collar just as Calgacus decided he had had enough, and leapt up to push the dog away from him.

Calgacus found himself with a mouthful of the poor woman's hand. She ended up with some bruising, and I was upset and felt guilty. The instructors were stressed and Calgacus was unable to join in with the rest of the class, refusing to even sit when asked, but laying at the side of the room, head down and tail still.

Ursula approached me at the end of the night to say "Try not to worry, but have you thought any more about getting Calgacus neutered?" Her tone was gentle and sympathetic; offering help.

"Yes," I said, my voice sounding weak and shaky, even to me, "but I've been waiting until Katie is more settled in before having it done."

"That is sensible. Maybe you would think of an injection in the meantime? It acts in the same way as neutering does, but if you find you don't like the effects it'll wear off." Ursula told me.

This sounded like a good idea. Ursula was so knowledgeable about dogs, and I trusted her implicitly.

"I'll ring the vet tomorrow." I told her.

# *9* Is it your hormones?

The injection that Ursula suggested contained the female hormone progesterone, which, at the time, was commonly used as a means of suppressing the male hormone testosterone, which drops dramatically in dogs once they have been neutered. Progesterone was used as a reversible way to test the effects that neutering may have on male dogs.

Neither Ursula, my vet, nor I had any idea that progesterone carried certain risks, so I wasn't particularly worried when, after having the injection, Calgacus became lethargic, didn't want to go on walks or run around, and stopped greeting visitors. The vet had mentioned that dogs sometimes became a little sleepy afterward, and Ursula talked often about how much dogs would calm down once neutered. I thought I was seeing a calmer dog. "Is he okay? He seems – I don't know – depressed, maybe?" I was often asked.

If this had been as far as it went, our story may have been very different, and also if I'd known just how badly wrong things could go. But it simply did not cross my mind that my big, gentle friend would ever become violent. There were signs, though, which I could have picked up on if I'd had a little more knowledge.

During this time, I took Calgacus to a local pet shop. We wandered around, looking at the food while I searched for treats that might cheer him up. We met a dog from our classes – a large, young dog – who was friendly with other dogs, and keen to greet Calgacus. This dog was playful and lively, and exactly the kind of animal that Calgacus tended to love. I took him over to her, thinking that he might be happy to meet a new dog friend. He sniffed her briefly, growled, then turned his head with the speed of a striking snake to snap at the air next to her head. This was something new – and very much unwanted behaviour.

I gasped his name and asked what was wrong with him? We returned home, with me feeling disappointed in him, and right there – without even being aware of it – I totally misinterpreted his feelings. My friend, who depended on me for everything, needed my help, and I let him down by concentrating more on his action being socially unacceptable than on how he might be feeling. I was cool toward him for the rest of the afternoon.

The next day brought more of the same. Calgacus, Katie, and I were walking together. She was still a puppy, under five months, and finding her feet in the world. She picked up a stick and was running in circles around Calgacus, trying to get him to chase her. He plodded along, head down, ignoring all her attempts. In the end, Katie dropped the stick and stood in front of Calgacus barking at him, trying to get his attention.

He turned suddenly, growling, and grabbed her ear. She yelped in fright and pain, and pulled away from him. When I looked at her ear, I found a small lump, the result of Calgacus pressing her soft flesh between his teeth. This – again – was unexpected: he'd never before bitten her. Once, when she'd wandered over and stuck her head in his food bowl, he had snarled at her, barged her, and chased her away from his food, but he hadn't touched her with his teeth.

I just did not understand. This wasn't like the problem we'd experienced with their meals: this was something different. It felt as though Calgacus was being mean, being so hard on a puppy who was being annoying. He could have simply pushed her away or ignored her: why did he have to bite her?

I marched him home from that walk and didn't talk to him for ages afterwards. Poor dog. He needed my help and support, and he didn't get it.  Even now, years later, I can't rid myself of the guilt of not realising that the injection might have played a part in these incidents.

These relatively minor events were the start, but they weren't enough to instigate a complete change, which actually occurred on a day that should have been a good one.

I had arranged to meet Nancy, a friend I'd known online for some time, and her cute little dog, Monty, and picked them up at the train station to take to my house. The first thing we did was let Katie and Monty into the garden together. The garden was bathed in sunshine, lighting the grass, the daisies and buttercups vibrant against it. Monty bounced on the grass, his cream fur fluffed around him, making him look like a small lamb. Katie ran around, tail wagging. She play bowed and scampered happily across the grass and into the bushes, trying her hardest to get Monty to chase her. He was so good with her; so playful and gentle with a puppy who was the same size as him.

Then I brought Calgacus into the garden, and, at first, things were okay. The dogs wandered about, sniffing and scent-marking. Something happened, and, before we knew it, Calgacus had grabbed Monty, lifting him off the ground. Monty screamed. I dashed to Calgacus, took his collar, and shouted at him to stop. He dropped Monty and allowed me to lead him into the house. Back in the garden, Monty was lying on the ground, bleeding. He couldn't weight-bear on one of his legs, and it was hard to determine how badly hurt he was. Blood dripped onto the ground, forming a puddle.

I was shaking and felt faint when I rang the vet, insisting that Monty be seen immediately. When we got him there, the vet told us to leave him, and said he would x-ray Monty in case of broken bones; would stitch him up, if needed, and phone to let us know when we could pick him up.

We returned to my house. It was surreal: Nancy and I had never met in person, yet here we were, dealing with the trauma of my dog having badly injured her dog. We sat at the table and waited for the vet to phone. I did what I could to help Nancy, bringing her tea and biscuits, and assuring her that I would cover Monty's medical expenses. When the vet eventually did ring, our anxiety increased.

## Canine aggression

"You can come and pick up Monty now," he told us.

"How is he?" We asked.

"I'll tell you when you get here," was his worrying reply.

His words sounded ominous. We got into the car, and, as I drove, it became obvious that Nancy was just as worried as I was. She turned to me and asked: "Surely the vet would have phoned and let us know if he'd had to do something drastic like amputate Monty's leg."

"I think so."

We fell silent until we were there. Thankfully, things weren't as bleak as we'd imagined. Monty had required extensive stitching to his torn muscle and skin, and needed lots of painkillers and antibiotics to prevent infection. He was delighted to see Nancy, lighting up when she came into view. The strong bond between them was evident then and for the entire time I knew them. The care and support that Nancy gave her dog always impressed me, and I feel deep awe for her dedication to the dog in her life. Monty had gone, with Nancy's care and love, from being an anxious little dog who'd barely left his breeder's property, and who had never been walked on a lead, to a happy little chap who lived in the middle of a city, and travelled all over the country by bus and train.

I drove home after driving Nancy and Monty to their home. The wheel in my hands and the road ahead seemed somehow unreal after the violence I'd seen, and when I went to bed, I lay long into the night with my eyes wide open, because every time I closed them, my mind was filled with images of Calgacus' jaws closing around Monty's back, lifting Monty off the ground, Monty screaming, Monty lifting his paw while blood dripped onto the concrete of the driveway.

Guilt weighed heavily on me, and both Nancy and I suffered flashbacks for weeks afterwards. I couldn't sleep and spent days in tears. Monty's wounds healed slowly, and he and Nancy had the added stress of coping with him developing a fear of dogs the same colour as Calgacus – a fear that, thankfully, left him once he'd had time to recover.

The world became a darker place for Calgacus and I as it became clear that he could no longer cope with other dogs. He stopped interacting with dogs that he knew and liked, including Katie, whose attempts to get him to play with her were ignored. He would bark and lunge at the end of his lead in an aggressive way to chase away any dog who was a stranger to him, and I have no doubt that, could he have reached the dog, he would have injured him, too.

Something I learned from this harrowing experience is never to accept advice at face value – not even from apparent experts. Listen, yes, but be cautious about following the advice without first checking the veracity of it – no matter who it comes from.

# 10 Let me give you some advice ...

I was now in a situation I'd never wanted to be in: living with a troubled dog –
one perfectly suited to the classes that Ursula ran for those with aggression
problems. And, instead of meeting people and chatting happily about dogs or the
weather, I was now the recipient of much harsh and wrong 'advice.'

"Make him so scared of you that he won't dare bark."

"Show him that it is unacceptable to behave like that."

"Carry a stick with you and hit him whenever he lunges at another dog –
he needs to know his place."

"My dog used to do that, too, and I carried a bottle of water and squirted
her with it when she did."

"Grab him by the sides of his jaw, drag him to you and scream 'NO!' right in
his face."

I knew that I could not do anything like that to Calgacus. He was still my
shining, striped puppy, and I couldn't bear the thought of having him be frightened
of me. But in addition to this, a quiet voice in my head pointed out that Calgacus
was a very large dog ...

I wonder now if that voice was a result of my dog-less childhood. I was
lucky enough to grow up with thoughtful parents, and their words were woven
through the fabric of my life. Dad often spoke about the risks of using threats and
intimidation to control the behaviour of others. As a young man, he was obliged
to spend some time doing National Service. Growing up in Glasgow, he was sent
from his home to an air force base somewhere in England, and told us how
friendly the people were he met there. How the others would leap up to make him
a cup of tea if he expressed a desire for one, or close open windows if he said that
he felt cold. He was impressed at the lengths his peers went to to make him feel
welcome and at ease so far from home, and felt lucky to have landed with such a
accommodating bunch.

Then one day he overheard two of the other men talking about him.

"Look out for Johnny Mac." Said one.

"Why?" The other asked.

"He's from Glasgow, and he shaves with a cut-throat razor."

"So?" Was the puzzled reply.

"Don't you know anything? He's from Glasgow. Bound to be in gang. Those
guys would slice you up as soon as look at you. He uses that razor to warn us.
Just do whatever he says. It's safest."

This was all news to my Dad, who'd never been part of a gang, and shaved
that way because he'd been taught it was the proper way. He hadn't intended to

# Canine aggression

threaten or coerce his peers but his normal shaving habits, coupled with where he was from, had that effect: they felt they had to stay on his good side or risk retribution.

No doubt some people would have been happy to be regarded this way; relished the status and control it gave them, but not my wise, gentle, sociable Dad, who understood the risks of inadvertently scaring his peers, who might decide that attack was the best form of defence. He posted home the razor and bought a packet of the safety razors that everybody else used.

I do understand the idea behind this form of deterrent. If Calgacus knew that every time he barked at another dog, he would be sprayed with water, I think he would have stopped doing so. But my refusal to try and frighten Calgacus aside, I was aware that his aggressive behaviour wasn't really the problem. Yes, it was a problem – a huge one for me – but it was a symptom and not the cause.

The root of the issue was that Calgacus no longer liked other dogs. He didn't want them close to him, didn't want to have to interact with them; didn't even want to watch them running around. His dislike of other dogs was the real problem, not the resultant behaviour. I wanted – more than anything – for Calgacus to like other dogs again, and no amount of water sprays would ever achieve that aim. And, of course, beyond not actually addressing the root of the problem, aversive tactics such as these carry other risks, which Behavioural scientist Murray Sidman covers at great length in his book *Coercion and its Fallout*. Using threats, violence or scare tactics as a means of influencing the behaviour of others runs the risk of negative fallout, which can be severe, and includes damage to relationships; the threatened individual sneaking around and hiding what they do; avoidance of all situations where the threats may appear, and violence directed at whoever is using these tactics.

Once Calgacus became aggressive toward other dogs, I read Murray Sidman's book from cover to cover, and I thought about my Dad's words. I used the knowledge as armour against the derision I received, and my own feelings of inadequacy, whenever I refused the advice of anybody who wanted to show me how to threaten Calgacus until he ignored other dogs.

There was no escaping the fact that I did need help, though, and so learning as much as I could about dogs became my new passion, and I took part in as many courses as I could – both practical and lecture-based. At first, I concentrated on watching how the experts I learned from behaved toward the dogs they were in contact with, and, over time, I also watched how they behaved with the people involved.  Those teachers were in an interesting position: needing to teach people how to teach their dogs, whilst also helping the dogs learn what they needed to.

Group dynamics are complicated. Some dogs are so keen to be close to other dogs that they can't focus if a playmate is too close. Others need distance, and can't focus if they see what they regard as the menace of another dog nearby. Others take reassurance from being close to a friend, and can only concentrate if they have the friend close by. People, too, differ in their ability to notice what is going on around them and respond to it – especially if their attention is taken up with learning a new skill. The canine students can't easily understand verbal language and the human students can't understand non-verbal language well. Layer upon layer of difficulty for anybody teaching in that situation: it's not an easy job.

Over time I noticed differences in the way that dog behaviour experts cope with these difficulties. Some spent time helping carers to understand their dog's emotional state, suggesting what it is they need to do to keep their dog safe, and providing tools with which to do this: space, things to hide behind; an acceptance that sometimes people may leave during group sessions and take their dog outside for a rest. Others seemed to have little patience for people, and were harsh – frightening, even – to anybody who struggled to understand or carry out instructions. And some were so enthusiastic about sharing their views that they would force them on those they were teaching, which, I noticed, made people reluctant to admit that dog behaviour can be more complicated than it might at first appear.

On one course, it was hammered home to me how effective this sort of control can be over a group, because I felt on edge and threatened by some of the comments. Early in the course, the instructor – a charismatic person with a wealth of experience, and huge expertise in teaching dogs and people – made a throwaway remark about dogs who bark and lunge at other dogs: something along the lines of

"It's easy to teach them not to do that, and anybody who has a dog who continues to do these things after a period of longer than a few weeks lacks competence."

Listening, I felt suddenly heavy, arms stuck on the table I sat at; feet welded to the floor. I looked down, taking in the pungent scent of the coffee next to me, then looked around the room. I wasn't alone in being still, staring at our coffee, or uneasily glancing around the room. I guessed I wasn't the only one with a dog who sometimes barked and lunged at things ...

More interesting to me now is that not a single person challenged this assessment. I remember us sitting there – a group comprising people who, like me, were fascinated by learning about dogs – feeling such collective shame that nobody spoke up or demurred about this condemnation. The power wielded by experts is huge.

I can't remember the reason now but, on one of the days, the instructor had brought a dog from her home – an elegant little dog who everybody fell a little in love with. At lunchtime, I walked with Calgacus around the park beside the training room, and saw the instructor and her dog come outside. The dog was on a lead, and they walked together, a picture of calm connection, moving toward Calgacus and I until the instructor looked up and saw us, and I caught the exact moment of her change in body language from relaxed to watchful.

43

# Canine aggression

The instructor stopped, stood still, and looked more intently at us.

I continued walking. I had no anxiety about the situation. Given her comment at the start of the course, I was confident that her dog would not bark at Calgacus, and I knew that, as long as I kept a distance from them, Calgacus would be able to pass a calm dog.

When we continued walking toward them, the instructor turned and went inside. The little dog had taken maybe only twenty steps before being taken back in.

I felt bad. Was this a comment about how Calgacus might behave? Had the instructor feared for the safety of the dog in her care?

I felt rattled and tense for the duration of that walk, and couldn't get what had happened out of my head. I spoke tentatively to one of the other students about it, and asked whether anything had been said about Calgacus posing a risk?

She laughed. "No. Nothing at all was said, but that *is* interesting. When you guys walked by and could be seen through the window, the little dog ran around, barking, and looking anxious. Do you think that maybe the cure for a barking, lunging dog isn't quite as easy as was made out?"

Now I feel grateful for those comments, and for that moment, which helped trigger my questioning of both my tendency to automatically obey those with authority and how people are instructed in situations such as this.

# 11 A decision to be made

There was still a decision to be made, despite struggling to cope with the changes in my life: when should Calgacus be neutered? The reasons for waiting – upheaval at home and coping with other changes – were stronger than ever.

Calgacus was dealing with his sudden dislike of other dogs. One of the greatest pleasures in his life up until then had been playing with canine friends, but, suddenly, this wasn't fun for him any more. Life was different for Katie, too, as her attempts to get Calgacus to play with her were rebuffed again and again.

I watched one night as she padded across the floor, all puppy pudginess and strawberry blonde fur, to where Calgacus lay. She bowed in front of him, front legs splayed out and bum and wagging tail high in the air. I sipped at my favourite thick, square, blue mug and tasted grassy camomile tea. Calgacus lay still, head resting on a raised part of the beanbag he was curled into. He sighed, and his ribcage expanded then contracted. Katie stopped wagging in front of him and tried to climb onto the beanbag to curl up next to him. Calgacus rose then, smoothly, his striped coat gleaming in the afternoon sunlight, and, without even a glance at Katie, padded to the sofa, climbed onto it, and lay down.

For me, too, nothing was the same. I knew what I wasn't prepared to try to fix things, but had no idea what I would do. These were all good reasons for waiting to let the dust settle.

Yet, there was now another aspect to this. I found myself thinking about all of the times another male dog had started a scuffle with Calgacus, even when he was happy and friendly, and worried that these might happen more often when he was hostile, which might make Calgacus feel even worse about other dogs. And I feared for the safety of any dog who got close enough to Calgacus for there to be contact. Even wearing a muzzle, Calgacus was big enough to pose a serious risk to smaller dogs. Might it be safer all round if he was neutered, and had less testosterone in his system? On balance, it seemed that having him neutered was the less risky option.

The aftermath of the operation provided an opportunity to build trust, and to practise slow, gentle handling with Calgacus. The operation itself went smoothly, and the wound healed without any problem. His stitches, however, worried me. As the swelling went down, it became clear that the stitches were so tight that the black thread was disappearing into Calgacus' pale skin. I looked for the knots, which would be slightly larger, bulbous, and more visible than single strands of thread, but could not find them. All I could see was slightly reddened, indented skin. When I gently touched Calgacus' inner thigh, with the intention of stretching the skin to see if it made a difference, he licked his lips, quickly and

# Canine aggression

often, stepping away as he did so, and looking at me, rolling his eyes so that I could see the whites (whale eye).

On the evening I took Calgacus to have his stitches removed, I spoke to the vet.

"His wound looks fine to me, but the stitches are very tight, and he isn't happy about being touched near them. Can you look and let me know if it all looks fine to you?"

"Yep – all looks good to me. Now, do you think he'll lie on his side while I get these out?" Was his confident reply.

"I doubt it. He doesn't even want me touching his inner thighs, never mind the stitches."

"We could get the nurses to hold him still."

"I'd rather not do that."

"Or sedate him so he'll stay still."

"Maybe. Or ... could you give me one of the tools that you use to take the stitches out, and I'll try to take them out myself?"

"If you think you can manage it. Remember, if he won't let you, we can always sedate him and do it here."

"That's what we'll do. I'll have a bash at home and if it doesn't work, I'll bring him back and you can sedate him." I concluded.

Off I went with Calgacus, stitches still in, and a little hooked blade. In most cases, having the vet remove stitches would be my preference, but Calgacus was keen not to be touched in this area. I knew that the vet would have limited time to do the job, and I worried that Calgacus would be frightened or hurt by the need to hold him still to remove the stitches. If I took them out myself at home, I could take my time, and make the process as easy as possible on Calgacus. I also had the backup of my vet if it turned out that I couldn't manage.

I'd come up with a plan during the day, and once we were in the house, I turned on the heating, as hot as it would go. In the kitchen I cooked a large pan of rice, which I let cool and then put in a dog bowl with some pieces of cooked chicken, and gave the whole lot to Calgacus. An hour later, I was sweating and drinking lots of water to ease my dry throat. Calgacus was lying on the couch, flat on his back, snoring.

Picking up the tool the vet had given me, I sat next to my dog and stroked him so that he knew I was there. I could see his wound more clearly as I stroked his inner thigh. There was a sharp intake of breath, and Calgacus abruptly opened his eyes, body suddenly tense.

"It's okay, sweet pea," I told him, "we'll do this slowly."

All the time I spoke, I was stroking Calgacus. He yawned, closed his eyes; relaxed. I took the little blade, gently stretching the skin until I could see some black thread. As gently as I could, I slipped the point of the blade under this, wiggling it until the thread snapped and I could pull it out. Calgacus leapt to his feet, pushing me aside, so I stood, went to the kitchen, and got some chicken. Calgacus followed me and I gave him some of this.

Then I waited until he was lying quietly again; this time on his side. I knelt on the floor in front of him, and propped his right hind leg on my shoulder so that I could repeat the procedure. He opened his eyes but didn't move, so I gave him more chicken. There were only three stitches to go, and I'd left the tightest one until last. This time, there were several failed attempts at wriggling the point of the tool under the stitch, and I had to pull gently on it, lifting the thread slightly to allow the tool to get underneath. Calgacus lay quietly throughout these attempts, waiting for the piece of chicken he received between each.

This wasn't a pleasant procedure for either of us, but taking the time to do it gently gave me an opportunity to build some trust with Calgacus. I'm sure that it stood us in good stead for the weeks to follow. Having him neutered turned out to be a good decision in terms of interactions with other dogs, as, throughout the rest of his life, Calgacus never suffered an attack from another male dog.

Visit Hubble and Hattie on the web: www.hubbleandhattie.com
hubbleandhattie.blogspot.co.uk
Details of all books • Special offers • Newsletter • New book news

# 12 Darkness and confusion

In the weeks and months after Calgacus' attack on Monty, I couldn't quiet the critical voice inside my head, and the walls of my life closed in around me, until it felt as if I was living in a small and isolated cabin. Previously, I had walked Calgacus and Katie three times a day – morning, lunchtime and evening – so we had lots of opportunity for social time and enjoying life together. Now, some days, I'm ashamed to say that the dogs didn't get walked at all, although they still had time to play in the garden, received attention, and were fed and cared for.

On one of the days where the yammering voice in my head was too strong to deny, I stood with my hand on the door handle, unable to move. I looked at my hand, and saw the fingers quivering as they rested on the old, brass handle. It was hard to get enough air into my lungs, and I breathed in short and fast gasps, as if I'd been running.

"Just go. It'll be fine" I said out loud, hoping that the sound of my own voice would be sufficient to make me take the steps to go outside and walk Calgacus.

Nothing changed. Still I stood there, hand trembling; breathing fast and shallow. My stomach was tight, the hard knot I could feel there suddenly tightening. I lifted my hand, turned away from the door, unclipped Calgacus' lead and went into the living room. This was too hard to deal with. I couldn't have a dog like Calgacus. What if a puppy ran over to say hello? Or Calgacus pulled the lead from my hand trying to get at another dog?

Leading up to this crisis, preparing for walks my mind showed me picture after picture of the risks outside. My body reacted, trembling, making it hard to breathe. But, as bad as that was, going out and walking could be worse, I felt. I'd anticipate the dog fights that would happen, and my stomach would churn as we stepped out the door, shoulders tense, anticipating trouble everywhere. I developed superstitions about walking routes, and would always cross the road at specific points so that I could avoid blind corners or crossings where dogs might suddenly appear. Returning from each walk, the sense of relief was so deep that I felt good for a time afterwards.

Calgacus didn't like any unfamiliar dog, but by far the worst for him were Spaniels. Their cheerful disposition, and tendency to run around quickly, changing direction frequently, and suddenly popping out from hedgerows seemed to be particularly problematic for him, resulting in a screaming fit, even if there was a Spaniel in the distance.

I found myself picking up on this. Driving to work or walking to the shops on my own, seeing a Spaniel running around in the park would literally ruin my day. I thought I was having a breakdown, and maybe I was, but I know now that my reactions were a normal response to trauma. Caring for a dog who is inappropriately aggressive is a difficult thing to do, and can result in confusion, upset, stress, and anxiety. The upset I felt is perfectly normal, and, in that, I wasn't alone – I just didn't know it at the time.

Before Calgacus was neutered, I had spoken to Ursula, my canine guru then. She specialised in working with dogs who had problems with aggression, and I pinned my hopes on her being able to help me get back my beloved companion.

"Calgacus attacked my friend's dog," I told her.

"Oh – a bit of a scuffle was it? Did you take him for that injection? That's the thing with un-neutered males." Ursula replied.

I sighed, "Yes, he's had the injection now. He's been in scuffles before. This wasn't like those times. This was worse. Monty – the dog he attacked – needed lots of stitches to hold muscle and skin together. The vet x-rayed him to check for broken bones. This was different."

I had her full attention then, and can still remember her eyes: large, blue; filled with sympathy and concern. When she spoke, she was tentative, wanting to help but also wanting me to know how serious this was.

"This isn't your fault." She said. "You've done all the right things with him. Sometimes these things just happen."

Then the blow. "I don't think I can help you." She said. "I don't think it would be safe to work with Calgacus in the way we need to with aggressive dogs. I think he'd attack you eventually – I've seen that happen before."

She was explicit. "He will never be okay with other dogs again. You can never trust him with them ever again. This is sad and it's awful because he's so young, but with dogs like him, once they've become aggressive, they will never be okay again."

She was as kind as she could be, and I guess that the situation was hard for her, too, as she wanted to help people and the dogs in their care live happier lives. I now vehemently disagree with the methods Ursula used, but she was definitely approaching this problem from a position of wanting to help. Yet here she was having to admit the limit of her knowledge: she couldn't help me. It must have been hard for her to admit. She chose her next words carefully.

"Some people with dogs as dangerous as Calgacus has become choose to have them put to sleep."

It's true, this does sometimes happen, but it wasn't something I could do. Not then.

Ursula understood. She gave me three pieces of excellent advice, which proved incredibly useful.

The first – teach Calgacus to wear a muzzle whenever we went for a walk – was useful once I got used to it. I was so ashamed, and expected everybody to run away when they saw us coming, but, as it turned out, I had absolutely no need to worry. Most people were lovely about it, and I was often stopped by

# Canine aggression

strangers, who would stand, patting Calgacus, and ask why he was muzzled. When I explained, to my surprise, they were supportive.

The muzzle was great for my peace of mind, as it meant that Calgacus was actually unable to bite other dogs. I still wanted to help him feel better, but at least I could walk him without flying into a blind panic at the sight of another dog. If his ability to bite was denied, I could relax a little while walking, and begin to look for opportunities for he and I to learn.

The second and third pieces of advice were both for Calgacus. Carry a pop-up umbrella with which to deter loose dogs, and walk in quiet, open areas where we weren't likely to come across many other dogs, and where there was room to escape should we do so.

Pop-up umbrellas are an excellent tool for protecting dogs who can't mix with others from being approached by off-lead dogs. We practised in the garden first so that I could learn how to handle the lead and umbrella, and Calgacus got lots of treats for ignoring the umbrella suddenly popping up next to him. Umbrellas are not things that dogs are familiar with seeing coming towards them, so most of them stop at the sight of one. Even if an oncoming dog doesn't stop, the umbrella can be used as a barrier to shield a dog who'd prefer not to say hi, until the oncoming dog can be removed, or becomes bored and goes away. It would also be possible to distract an approaching dog by throwing handfuls of food from behind the umbrella, making an escape while they were eating.

Walking in quiet, open areas is also something I'm enthusiastic about for dogs who dislike strange dogs, as to do so can encourage relaxation, and promote good feelings. Stress in both dogs and people causes a number of physiological changes that are designed to help the individual deal with the situation – the fight or flight response – which means that a dog is likely to exhibit behaviours associated with strong emotions, such as aggression or running away. They are also less likely to be able to learn or respond to requests from their carer.

Not allowing sufficient time to recover after a stressful event means the effects can become cumulative, making it harder, subsequently, for a dog to respond in any other way than with aggression or by running away. Keeping stress levels low is essential if worried dogs are to learn to relax, and feel better about being around other dogs.

Ursula's suggestions helped a lot, and most days I did manage to quieten the worrisome voice inside enough to walk Calgacus and Katie – often one at a time so that I could concentrate properly on each dog. My walks with Calgacus were filled with sadness, as I couldn't help but compare the dog he'd become with the dog he used to be, and sometimes those who'd also known him before expressed similar sentiments.

One of Calgacus' friends was a  pretty black, tan and white Collie cross. She was slender, with long legs and beautiful pricked ears, and she loved other dogs. Before he developed problems, Calgacus and this dog had loved each other. They would play chase games – with Calgacus trying desperately to keep up – and sometimes they would wrestle. He was always so happy to see her.

The first time we met each other after the attack on Monty, the other dog's carer asked why Calgacus was wearing a muzzle, and why he was on a lead.

"He really doesn't like other dogs," I told her.

"But he loves them. He loves playing so much," she replied, looking at me oddly.

Her dog play bowed to Calgacus and ran away a few steps to entice him to chase.

"Please let Calgacus off his lead so that they can play," She said.

"Okay – but I don't think he will play," I told her.

We were in an open space, and I could see that there was nobody else in the field, so I unclipped the lead and whispered to Calgacus that he could go and play. His friend ran over, nudged him and ran away on fast legs, her eyes sparkling in anticipation of a game.

Calgacus turned his head away, looking over his shoulder, not engaging with his friend at all. He looked sad and defeated, and my heart broke for him. The Collie cross tried again with the same result, and then again, before she gave up and came to stand next to her carer.

"I see what you mean. I didn't believe you," she said.

"I know. I wouldn't have believed it either. Calgacus loved playing with other dogs so much." I said.

We parted company, and I put Calgacus on his lead again before we moved off, walking together; both of us, I'm sure, feeling a great sadness. Things seemed hopeless.

All I could think about was getting Calgacus back to how he'd been. I shut out what Ursula had said about how he never would be, and I made myself carry on taking Calgacus to classes. I know that this was not a kind thing to do. He wasn't okay around other dogs, and being forced into a room full of them upset him. It wasn't a kind thing to do to the other dogs, either: they didn't deserve to have Calgacus barking at them. But, for me, the classes had become something of a life-line. I lacked the necessary skills and experience to deal with Calgacus' problem; I didn't have friends with dogs, and my family had never had dogs. The dog club was the only source of help and support I had access to at the time, so I carried on taking Calgacus to class.

But, as bad as things felt for me, they were far worse for Calgacus. I believe that, when he attacked Monty, Calgacus was feeling depressed and frustrated, and lashed out because of this. I suspect that, afterwards, he felt ashamed and distrustful of what he might do, as I would do if my own behaviour deteriorated as badly as that: frightened to go to meetings and possibly even to be around people in case the same thing happened. I know that this is to anthropomorphise his behaviour, but I believe that Calgacus struggled to be around other dogs because he was afraid of something similar happening. Being in the close confines of the

# Canine aggression

hall with other dogs during class was something that I now believe terrified him.

If I could reach into the past and speak to the woman I was then, I would advise her to stop taking Calgacus to classes immediately, to seek help and support in other ways, and to let Calgacus relax and begin to recover from the emotional trauma he had suffered. But, because I didn't know any better, I persisted in taking him to class. In busy classes, he spent so much time barking and lunging at the other dogs that even I didn't think this was helpful for him, and I decided to move to one of the club's quiet classes to see if that would help. We turned up to a class with just two other dogs: a female Greyhound and a female Pointer: both quiet, gentle dogs who tended to move slowly, and were not at all confrontational with other dogs. Neither of them even looked at Calgacus.

I sat as far away from them in the hall as possible, and tried my best to focus on Calgacus, to engage with him, and to use his favourite food to encourage him to ignore the other two but it didn't work: he simply *had* to know what the other dogs were doing.

At one point in the class, the Greyhound and her carer were practising an exercise, which entailed the dog being at a distance from her, and then running or walking to her carer. The instructor was supervising them, and the situation was safe.

The Greyhound wasn't near us, and wouldn't have been able to get to us, yet, Calgacus couldn't deal with this, and every time the Greyhound took even a step toward her carer, he would leap up, fly to the end of his lead, and roar out a warning to her. She got the message and stood still, too frightened to move another step.

I had been trying to distract Calgacus with food, and encourage him to turn to face me, but without success, and I was tired and upset. When the Greyhound froze to the spot in fear, I looked at Calgacus, and it was as if I was seeing him for the first time. His distress was so evident that the air around him was heavy with it. For the first time I appreciated just how upset he was about having other dogs around him. He wasn't misbehaving or ignoring me; my poor boy was in a state of complete panic. He could not allow that dog to move because, if she did, she could come over to him, and something terrible might happen. Calgacus was dealing with the situation in the only way he knew how to – by making sure that the Greyhound did not move.

It was obvious, then, that Calgacus couldn't attend class any longer, given the trauma I was causing him and the other dogs. The club couldn't help, and I had nobody else I could turn to.

Although a low moment for me, the events that night did me a huge favour, because, for the first time, I really noticed Calgacus distress and did something about it. I cared deeply about Calgacus, and desperately wanted him to enjoy life again, but what I needed to learn was that care and love weren't enough. I had to take action to help him feel better and this included not taking him to the dog club any more – even though I felt I needed the support I felt it provided.

# 13 Flashes of insight

It would be good to mention at this point some of the insight and guidance I received from people who were not dog trainers.

A couple of ideas that initially made sense to me I subsequently abandoned (the rank reduction book I read before Calgacus moved in, and the way that aggressive behaviour was dealt with in Ursula's training classes) because messages I received from reading, and listening to others not directly connected to canines, allowed me to shape my thoughts around how I wanted to help Calgacus. As already noted, no end of people are willing to offer advice on the subject of problematic dogs, some of which – like Ursula's umbrella idea – is excellent, and some of which – like scaring the dog into dropping undesirable behaviour – is both unkind and cruel, and has the potential to make matters much worse.

Finding a way to navigate the often conflicting advice is tricky, but the following describes some of the things that helped me find my way with Calgacus.

Reading about Stanley Milgram's work was pivotal, and started me thinking about what being an authority figure for my dogs might mean, and ultimately helped give me the confidence to look for ways that the dogs and I could get along, without me becoming the sort of authority figure who must be obeyed without question.

Stanley Milgram was a psychologist studying human reactions to authority during the 1970s. The first edition of the book he wrote about his experiments, *Obedience to Authority: An Experimental View*, was published in 1974, the year after I was born, and Milgram died in 1984, when I was eleven. His words and research have had considerable influence on my thinking, forcing me to reconsider some of my deeply-held beliefs; giving me hope by so doing. I like to think that I am a kind, compassionate, empathetic person who lives by these tenets, but Stanley Milgram's work made me realise that I cannot always count on myself to do so. My autonomy, and thus ability to behave in ways I like, is more limited than I imagined.

Milgram's experiments were simple, and considered to what degree people would obey a stranger who appeared to be an authority figure (a scientist wearing a white coat, apparently carrying out an important study). My feeling was that probably nobody would be obedient, because they had no interest or investment in the outcome of the study, and their jobs and career progression didn't depend on them doing well in these experiments. The authority figure had no control at all over their lives, and, even had he wanted to, could not inflict any negative consequences if they had refused to comply with his instructions. Under

# Canine aggression

those circumstances, people would simply refuse to obey any request they weren't happy about, I felt.

I was chilled to read that the results did not bear out my hypothesis.

The volunteers in Milgram's experiment were ordinary people making their way in the world, yet they obeyed instructions from the scientist that they believed would result in hurting another person, who, as far as they knew, was just like them. And most of the volunteers continued to follow the instructions even when they could hear the victim apparently screaming in pain, and begging them to stop.

Of course, nobody was really harmed in this experiment, although the volunteers believed that this was the case, and carried out the instructions because a calm, authoritative figure in a white coat told them it was important that they did so, and that he would take responsibility for any negative outcomes. All of them were free to walk out whenever they wanted to, yet most of them did not. Such is the strength of our obedience to authority that most of us will do what we are told, even if we believe that it is not the right thing to do.

Studying these experiments provided two things. The first was a conviction that finding ways to dilute my own authority, whilst trying to increase the freedom of those around me was a worthwhile thing to do. Dogs (and all animals), I'm sure, live richer, happier lives if they feel safe and comfortable about exploring and questioning the world around them.

When it comes to people, doing what I can to make it feel safe for others to question authority is one of the most useful things I can do when I teach, particularly when it comes to teaching about dogs. I am committed to using kind methods to do this, and to continuing education about how to do this well.

The second realisation I made from reading Stanley Milgram's work, and talking to others concerned dog training classes, and how natural it feels for participants to bow to the instructor's authority, even if it means doing something to their dog, or asking something of him, that they feel badly about afterwards. In order to help Calgacus I needed to seek advice from experts, but wanted to be as sure as I could that he and I would be safe from poor and bad advice.

In this respect, some really good things can be done, and one of the easiest is to ask in advance about the instructor's methods. Follow this up with a request to watch the instructor at work with another client or set of clients, so that you can actually see how he or she teaches, and what happens in class, before taking your dog along.

One of the most life-changing and powerful methods that an instructor can employ is to encourage a questioning, reflective attitude in those they seek to teach. Allowing others the space and safety to say "Can you explain further?" or "I don't think that'll work for us" is far more important, in my opinion, than having people just do what I want them to, even if I'm certain I have the knowledge and answers to help them, and really want them to listen to me.

If ever I am in a situation where I am the teacher, the authority figure, I am thrilled if any of those I am teaching are willing to discuss with me why things might not work for them, or to express scepticism, and even outright refusal to do as I ask. I like to encourage questioning and self-aware behaviour, and for others to see this happen, because when we feel able to challenge authority, we are more likely to resist a situation in which our dogs may be scared or even hurt in a training class.

The amazing animal trainer, Ken Ramirez, was another inspiration, who I travelled to see in person at a three-day lecture attended by many from across Europe. I sat for three days in a huge lecture theatre and listened. What I heard filled me with a mixture of butterfly-in-stomach excitement and hope for the future that was full of optimism and enthusiasm. The basis of Ken's message dealt with changes and rewards.

I'm sure I'm not alone in wishing that I could resolve problems using the same approach and methods that I always have, but, as Ramirez eloquently pointed out, that simply isn't possible. To see changes in the behaviour of animals, or indeed, myself, requires that I change.

What was different about this lecture was how practical it was. Previously, I'd found these sorts of discussions tricky, and often felt defensive about all the things I wasn't willing to change. Ken Ramirez talked about making lists of things we might change, prioritising them so that the most important ones were evident, and then considering ways to tweak and subtly alter our daily life to allow for small changes that might build toward the most important things. This could be something as simple as deciding to spend five minutes of each dog walk working on loose lead walking.

The other strategy which has stuck in my mind is rewarding desired behaviour. I have noticed that class participants very often want to know when they can stop rewarding their dog for desirable behaviour; at what point will their dog do as they're asked for no other reason than that the behaviour has been requested? The point that Ken Ramirez made was that, surely, in order to build and maintain good relationships, it makes sense to continue to do nice things for those we care about? Lots of married people find themselves in the divorce courts when they forget that doing nice things for each other is important, even after the relationship is established (with probably more reason to be nice!). Lots of people who care for dogs find that once they stop making an effort to reward their dog's good behaviour, the dog stops making an effort to behave in this way! As Ken explained his ideas, they made perfect sense to me, and it was good to hear him say that he does not, at any time, phase out food rewards for something that the animal has learnt well.

I was travelling to visit friends, and had some time to spend in Edinburgh airport while waiting for my flight, so gravitated toward the bookshop in the airport. I've always found these to be magical places, filled with different worlds, and, for me, as yet undiscovered knowledge. From the enticing array of titles on display, I picked up a book called *The Lucifer Effect*, written by Philip Zimbardo,

# Canine aggression

within which was a rich source of contemplation that seemed to apply to my life. The book – an account of an experiment conducted by the author at the university in which he worked – made it clear how strongly environment matters, for all species: where we live, who we come into contact with, how we are taught, it all matters. The structure of our worlds and what we experience shapes us, and re-shapes us after change. This book made me uncomfortably aware of my vulnerability to becoming abusive towards my dogs, and turning them into the depressed, oppressed subjects of a tyrant. I realised that I have no special powers or inner self that would prevent this from happening, and that a change in my environment could dramatically change how I behave.

I began to look for ways to set myself up to be able to follow the path I wanted to be on, and now totally avoid activities where people feel they must shout at, jerk on the lead, spray with water or otherwise hurt or scare their dogs into doing what they want. I also totally avoid training groups where the teacher is unkind to the human students.

Over time, I've increased my attempts to create a good environment for myself and my dogs, and now seek out those trainers and behaviourists who reject hierarchies and unjustified authority, and who prefer that their students – and their dogs – have a voice, and are heard. Reading *The Lucifer Effect* helped me to realise the necessity of doing this, and, furthermore, helped me become much more intentional and careful about the people I regularly spend time with; seeking social groups that support the way I want to live, and letting go of connections that are at odds with the way I want to go about things. Reading Philip Zimbardo's work was life-changing for me.

The importance of environment also came up in a book that looks at the world from a different perspective. Bruce Lipton's *Biology of Belief* covers the field of epigenetics (the study of how inherited genetic characteristics are influenced by internal and external living environments and experiences), and I was surprised to learn from this that none of us is completely 'fixed' by our genes, as I was taught in biology class at school. It seems that the world of biology has moved on in the decades since ...

A part of the book that stuck with me was a description of a study by Robert Waterland PhD and Randy Jirtle PhD titled 'Transposable Elements: Targets for Early Nutritional Effects on Epigenetic Gene Regulation,' originally published in 2003 in the journal *Molecular and Cellular Biology*, which dealt with the influence that diet in pregnant mice can have on their offspring.

I read that part of the book over and over again, and found the actual study on-line to read – although much of it was above my head. I've been repeatedly taught over the years that the traits and characteristics of individuals are fixed by inherited genetics at the moment of conception. Yet, here was a study that said 'not always' to that. My world shifted when I read that chapter. That the environment inside the womb could so profoundly affect baby mice gave me much to think about, and was further evidence that there is little about any species that is absolutely fixed. Habits are, without doubt, hard to break, and each of us is inherently predisposed to certain characteristics. But we can, if given the opportunity to alter the environment around and inside us, make some changes.

These learning opportunities have helped set me on a path to look for ways to empower the dogs and people around me; to learn how to let go of the need to be in charge, and to spend time and energy with those who will help me learn what I need to in order to live the life I want. One of the things I routinely wrote about as I went through the process of working through problems with Calgacus was my own anxiety levels.

Here are some of the things I noted –

• From 2005
I make things more difficult for myself by worrying. Sometimes, even before I go out, I'm already on edge because I've imagined the worst case scenario before I've even set off.

• From 2007
Since Calgacus' problems began, I have had a real problem with worrying about dogs behaving aggressively toward each other. Watching dogs close to each other – particularly in enclosed spaces as they would be at workshops – causes me huge amounts of anxiety.

From 2008 on, though, I found that I reported feeling better and making progress with this; saying things like: "I was pleased to note that my anxiety over potential aggressive incidents was very low."

I suspect this came about partly with time, partly through spending time with people who were calm and relaxed, and partly by accepting I had a problem and needed to find ways to deal with it. I'm happy to say that improvements continue, and, even though I still feel it is a good idea to be cautious about letting dogs meet one another, I can be in rooms where dogs do this without feeling the need to leave or move away. I no longer consider that dogs are a threat to one other. Attacks such as the one Calgacus carried out on Monty are rare events, thankfully.

# 14 A small setback

Losing the support of the dog club was difficult, even though it had become clear that carrying on with this was making things worse. I had to find another source of help.

I'd heard about Tellington TTouch from an internet dog forum, and from Ursula, who had always been enthusiastic about it, and would sometimes incorporate it into the work she did with troubled dogs. I wasn't clear on exactly what this method was, so the first time I really saw it in action was on television. When the series *Talking to Animals* aired in 2003, my partner called me into to the living room.

"Come and watch this. This woman is doing amazing things with dogs," he said – or something along those lines.

I sat in the living room with him and watched for a short time.

Then I stood. "What a load of wishy-washy nonsense," I said, "there is no way that could help Calgacus."

I stamped out of the living room and went back to the kitchen, where I added more cheese to that night's meal, and looked around to see if there was any chocolate in the house. Inside I was a turmoil of emotion: anger at the suggestion that there could be an easy solution to the nightmare I was living, and disbelief that something as gentle as TTouch could actually help.

Having dismissed TTouch, I went on to spend a weekend doing something that could legitimately be described as much more wishy-washy. I'd read an article about Emotional Freedom Technique (EFT) in *Dogs Today* magazine, which, based on acupuncture, uses tapping with the fingers on pressure points rather than needles. Maybe what attracted me so much to this particular therapy was the article's claim that it can help both animals and people to overcome emotional problems, and I hoped I could learn something that would help Calgacus and I.

I booked on an EFT weekend course which took place in a large mansion house in rural Scotland. That weekend the sun shone, making the garden flowers glow with colour. The grass in the garden was short, and so well kept that it resembled a green carpet. Inside, I was allocated a room with a large bed covered in a pink, rose-patterned bedspread. A comfortable armchair stood in the corner of the room. Meals were prepared for us, and breaks were spent walking in the estate grounds. Nothing more could have been done to make the weekend a relaxing one.

Despite the amenable surroundings,unfortunately, I didn't find EFT useful for Calgacus or my own emotional problems, and an event that weekend brought home to me just how much I was struggling.

On one of our walks we met a small dog who lived on the estate. He and the three dogs who belonged to the course instructor were friends, and the dogs greeted each other with softly wagging tails. Somebody from the group bent down to pat the new dog while the greetings were going on. I looked on and felt my heart race, as my face got hot and I felt weak and sick. I was convinced that one or other of the dogs would become angry and turn the greeting into a violent attack. I wanted to push the kneeling woman, and scream at her to get away from the dogs.

Nobody else was bothered: those with other dogs were unconcerned, and everybody was smiling and talking to the dogs. It felt like I was in a nightmare: all I could see in my mind's eye was dogs grabbing each other, people screaming – and blood. In the end, I turned away and pretended to be admiring a tree, taking shaky breaths and hiding my trembling hands.

Once I was home, I noticed more problems in this respect. I didn't like watching dogs greet each other in the park, even if they were dogs I'd never met and had no emotional involvement with. When I saw dogs in groups, I hated it if any were close to each other, holding my breath and squeezing my hands into painful fists if I saw the dogs sniffing each other.

While I didn't find EFT useful for my or Calgacus' problems, it's pertinent to mention that I did once have a fascinating experience with it. At the time I did the course, my friend, Hilary, and I were volunteers at a local rescue centre, where we'd spend a few hours each Saturday walking and training the dogs.

One of the dogs living there for a time was a terrier cross. He had a sleek, black coat over most of his body, and his legs – which were shorter than might be expected, given the size of his head and body – were a mottled mixture of rich brown and black. This dog found life difficult. If anybody approached his kennel, he would growl and dash to the outside run to get away. The staff told us that, if cornered, he would bite.

Such was this little dog's fear that, even after he'd been there for a few weeks, Hilary and I hadn't been able to gain his confidence sufficiently to take him out of the kennel. On the day in question, Hilary spent time in his kennel with him. She didn't look at him and, instead, stood still, dropping food on the floor and speaking to him in a soft voice. The dog would eat food that had rolled a bit away from her feet, but wouldn't come any closer to her. After a while of this, Hilary came out of the kennel, and we gave up on the idea of walking him that day. We hoped – as we had done the previous weeks – that he'd feel able to come for a walk the following week.

There was a letter pinned to the front of the dog's kennel from a child in the family which had given him up. In it, he was described as loving cuddles and having his ears scratched; he would love anybody, the letter said, who was willing to play tug games with him. I stood outside the kennel looking in at a terrified, growly, little dog, comparing him with the description written by somebody who had loved him. Thinking of just how much confusion, fear

# Canine aggression

and trauma the dog must be feeling, following the loss of his family, and his imprisonment in an unfamiliar and stressful environment, brought a heavy lump of sadness to my stomach.

I decided that I would spend some time in the little terrier's kennel with him, and try some EFT. One of the things I'd been taught on the weekend course was that, when working with animals, the practitioner can tap themselves whilst thinking about the animal's problem, which allowed me to help the little dog without touching him, or even being close to him.

After the first round of tapping, he would eat out of my hand, as long as I lowered it to his level and made sure I didn't look at him, and another round of tapping concluded with him jumping up at me to try and get more food! I looked down at him and his eyes were different: softer and brighter than they had been, and the stress wrinkles across his forehead and around his mouth were gone, too.

I crouched down next to him without thinking, and felt stupid. This dog had behaved aggressively toward everybody he'd been in contact with at the centre, yet I had crouched down so that he could bite me in the face if he became scared. What was I thinking of? The little dog put his front paws on my knees, reached up and licked my face.

There was a toy in his kennel so we had a game of tug with it, and, as we played, I noticed a couple approaching the kennel. Standing up, I expected the little dog to growl and run outside as usual.

"He's nervous," I explained, "scared of strangers."

The couple looked at me without saying a word, and then they looked down... at the dog who had his front paws up on the bars of his kennel, and was looking at them and wagging his tail.

He found a home quite quickly after that and I never saw him again. I think about him sometimes, and wonder what made the difference. Was it the EFT? Habituation to our spending calm time with him? Did the EFT make a difference to my posture and breathing that helped? Had something changed about me once I read the note and reflected on how horrible things were for the dog? I just don't know, but, whatever it was, it helped, and although I can't remember the last time I tried to use EFT, I keep the book from the course in case I ever want to remind myself about it.

Eventually, having had a long break from classes, I began to look for new sources of expert help. I knew that my skills weren't up to the huge task of helping Calgacus feel better about other dogs, so I looked for help.

My first port of call was a behaviourist who belonged to a UK-based organisation that specialised in helping those who struggle with their dogs. Wanting qualification-backed assistance, I looked for an organisation whose members had degree-level qualifications in dog behaviour. I emailed, explaining what had happened with Monty, and the new problem that Calgacus had with other dogs. Then I waited ... and waited ... and waited. No reply.

It may have been that the individual concerned was simply too busy to take on anyone new, or maybe it was because of Calgacus' size and breed. Ursula's opinion had been that the problem was not fixable because of the sort of dog Calgacus was; his genetics determining that the problem would have occurred, regardless of what I did. And she wasn't the only one to tell me that: years later, when I joined an online forum for Bull Mastiff enthusiasts, their view was that, if the breed (and males in particular) has a problem with other dogs, this can never be resolved.

Or maybe the lack of a reply was due to worries about safety and/or legality. Working with aggression in dogs carries a big responsibility, which obviously increases according to dog size. The answer remains a mystery, and, after a while, I looked elsewhere. This time, I did what many people do, and asked one of the veterinary nurses at the practise I use for a personal recommendation. She gave me the details of a nearby class which, she said, she'd heard really good things about. What was to follow was a powerful learning experience that showed me that not all dog classes are good places to be. The process began with a phone conversation with the class instructor, which went like this –

"I'm looking for help with my Bull Mastiff. He's become aggressive toward other dogs. He bit my friend's dog and injured him pretty badly, and I believe that he might do the same thing to any dog he got to," I said.

"How is he with people?"

"He's fine with people. The problem is with other dogs."

"Are you sure? These dogs are often a problem with people, too. I used to have one in my classes, and it would sometimes bite its owner's arm."

"I'm sure. He's fine with people."

"Okay. We have a busy puppy and beginners' class at 6pm on a Thursday. Bring him down next week and I'll have a look at him."

"I'm not sure that's a good idea. He really doesn't like other dogs, and I'd hate for him to scare somebody's puppy," I replied.

"True. Well, you could muzzle him and then he won't be able to do any harm. Will he wear a muzzle?"

"Yes, he will. The thing is that he can be very frightening. The last class I took him to had just two calm dogs in it, and he scared them both so much that one of them wouldn't move at all. It would be very bad for puppies to be in the same room as him."

"Maybe you're right. Bring him along to the advanced class at 8pm. There are fewer dogs, they're all adults, and all under good control. Now – what do you walk him with?"

"He wears a head collar for walks."

"A choke chain would be better. Give you more control."

"Hmm – I'm not sure," I replied, "I like that the head collar gives me control of his mouth. He won't bite another dog using his neck, and so I'd prefer if I had the ability to control the biting part of him."

# Canine aggression

"Maybe you're right. Yep – stick with what you have and we'll see you at 8pm on Thursday night."

The red flags in that conversation are very obvious, now. Here was a professional instructor, willing to risk the psychological well-being of the puppies in his class, going on to suggest I use a piece of equipment designed to cut off a dog's air supply when used, and at a time when the dog in question is considerably stressed. I should have known better but I didn't, despite all that had gone before.

Calgacus and I turned up to class the following week. This class took place in a huge hall, and comprised only a few dogs: well behaved, sensible dogs, who wouldn't run over to say hello.

"I'll show you how to do heel work," the instructor said, holding out his hand for Calgacus' lead. "Will he eat sausages?"

"He'll pretty much eat anything." I told him.

"Right, then. Hold a sausage here, right in front of his nose. Keep your hand by your hip. Don't let him grab the sausage. Walk in a straight line. Off we go."

Calgacus looked beautiful, trotting along, his feet placing smoothly and his legs moving with oiled grace. His striped coat gleamed, and he gazed attentively at the sausage in the instructor's hand, his head never leaving contact with the instructor's left thigh. The two of them walked in straight lines, turned at right angles, turned completely around, turned to the left, turned to the right, and all the time, Calgacus stayed in place, his head touching the instructor's leg whilst a steady stream of silvery gloop drooled from the sides of his mouth and onto the instructor's trousers.

After a few minutes, they stopped and Calgacus was given the sausage he'd been following. The instructor handed the lead to me and, as I took it, he looked at his trousers.

"LOOK AT THE STATE OF ME!" He cried, brushing at his clothes, and getting gloopy mess all over his hand.

I assumed he was joking and laughed. "That's why I don't encourage him to touch me when we walk."

"THIS IS NOT ON! I'LL HAVE TO WASH THESE WHEN I GET BACK." He continued to shriek.

Calgacus stood between the instructor and I, looking longingly at his treat bag, waiting for more sausage, and ignoring the shouting, which I was becoming increasingly aware wasn't a joke. Realising that more sausage was not forthcoming, Calgacus firstly glanced at me and then dipped his head. Understanding what this meant, I stepped back so that I was as far away from Calgacus as the lead would comfortably allow, whilst the instructor stepped into the vacated space and continued to berate me. He wasn't looking at Calgacus, and, even if he had been, he probably wouldn't have realised what the action indicated.

In movements so careful it was almost in slow motion, Calgacus shook, long and enthusiastically, rotating his body from side to side, the movement going from his head all the way to the tip of his tail. Slobber flew from his lips in huge gobs, covering the instructor's T-shirt, head, arms and legs in a shower of drool. I laughed – I couldn't help myself! We were told to sit down after that and watch the others do heel work so that we got the idea.

"You have a go now," the instructor said, looking at me. He indicated that Calgacus and I should come up on our own to practise. I was nervous, concentrating more on Calgacus than on where I was going, trying my best to keep his attention on me and not on the other dogs.

"Walk in a straight line," the instruction boomed into the room.

I stopped, red-faced, butterflies flitting in my stomach, a trembling hand holding the tube of cheese I was using to reward Calgacus. "I'll try."

Another few seconds passed and then I felt hands on my shoulders. I stopped walking, slightly shocked. "I said – WALK IN A STRAIGHT LINE," the instructor's voice was loud in my ear. He gripped my shoulders and shook me back and forth as he spoke.

I was passive, in shock at this treatment. Calgacus stood next to me, waiting, probably looking at the other dogs and assessing how much of a threat they were, or wondering if there might be a sausage in it for him. I say probably because I have no idea what he was up to – I was too shocked by the assault the instructor was subjecting me to. Nobody else in the room seemed bothered, and sat watching or chatting together. Presumably, bullying and physical intimidation had become so regular in class that it was barely worth a glance.

We left that night and didn't go back: chalked it up as something else that didn't work out. I might not have appreciated at the time that this was what I did, because I hated what was happening. I hated not knowing what to do; the struggle of trying to find help, and that so many people wanted to write off my young and healthy dog as broken and beyond help. With the benefit of hindsight I can appreciate what a time of rich learning this was, when I was able to try on new things for size and move on when they didn't fit. I experienced first-hand just how difficult it is to get help with a dog who has behavioural problems.

And within it all, even given the difficulties, there were some lighter moments. I can still see Calgacus shaking slobber all over an angry instructor, which was one of his many attention-seeking tricks. That he felt able to do this in such a tense situation still makes me smile.

# 15 Finding direction

I don't know what drew me back to TTouch. Maybe the short segment of *Talking to Animals* that I'd watched had stayed in my mind; maybe I was feeling particularly disheartened by failing, thus far, to find help, or maybe I read something on one of the forums I frequented that gave me hope. Whichever of these – or a combination of them – it might have been, I found myself on the internet one day searching for books on TTouch, and discovered *Getting in TTouch with your dog: a gentle approach to influencing behaviour, health, and performance* by Linda Tellington-Jones, which I heartily recommend, along with *Unlock your dog's potential: how to achieve a calm and happy canine* by Sarah Fisher.

I read the book, my state of mind a mix of irritation and hope. I looked at photos of a dog learning to be calm around other dogs, and I felt a soft butterfly flutter of expectation in my stomach. I flicked through a few more pages, the glossy paper slippery under my fingers, then settled down to read the text, looking for advice that might help dogs who were upset by their own kind.

The method involved using my fingers to move the skin of Calgacus' shoulders, head, muzzle and lips in one-and-a-quarter (on a clock face, from 6 round to 6 and then on to 9) circular movements. To gently stroke his ears between my thumb and forefinger. To move his tail in circles, and gently agitate each of the little joints between the vertebrae. Every night I sat on the couch and performed these exercises with him.

Calgacus was stoical about my attempts: he seemed to understand that it was important to me, so stayed where he was, co-operating when I asked him to turn over so that I could work on each shoulder in turn. An intense, stressed kind of concentration seemed to prevail whenever I approached Calgacus to do his nightly TTouch sessions: I badly wanted it to help but couldn't help feeling that I was clutching at straws.

About two weeks after the book arrived I was out walking with Calgacus in a small field close to our house. The weather was beautiful – one of those perfect spring days in May when the world seems full of promise.

We rounded a corner, and ahead of us on the path was one of Calgacus' old friends, a handsome Golden Retriever coming toward us with his carer. Initially, I didn't look at Calgacus, expecting that he would be ignoring his friend as he usually did in those days, but smiled and said hello to both person and dog. Then, glancing down at Calgacus, I saw that he was sniffing the Retriever's nose, his tail gently wagging: a small movement, quiet and tentative, but a wag, nonetheless.

I went home, wondering about the TTouch. Could it have made some difference? I thought it likely it was having some kind of effect. Calgacus had been ignoring his dog friends for months at that point, and the only thing that had changed was our using TTouch. That evening, I sat at the computer and, with the window flaring red in the sunset, went online to find out more about TTouch, discovering, initially, information about how to train as a TTouch practitioner. I sent an email inquiring about this. Then, as the sun slipped lower, leaving the window a deep, blue square beside me, I found a day event nearby. Half a day on TTouch and half a day on something else I'd never heard of: something called 'calming signals' ...

This phrase was coined by the well known Norwegian dog expert and trainer Turid Rugaas, who noticed that dogs seem to have ways of communicating with each other intended to prevent conflicts becoming physical and potentially injurious. Like many who spend much time with dogs, Rugaas believes they want to avoid physical confrontation as much as possible to avert the risk of injury.

The presenter of that half-day workshop spoke well, and was interesting and engaging. The dogs she brought along with her were able to demonstrate some of what she was talking about. I remember being uncomfortable about watching the dogs freely wander around the room, mixing with each other and stopping sometimes to be stroked by those attending. I moved to the back of the group to stay out of the way, and my friend, Hilary, joined me.

"Calgacus doesn't do that," I said to her, leaning closer to whisper.

My arms were folded and I sat back, not at all receptive to the idea that Calgacus might try and avoid violence, but quite the opposite, I felt, as violence seemed to be his preferred option.

That talk, like so many of the new things I discovered, rolled around in my brain, popping up at different times, helping me understand new notions. I noticed that Calgacus actually did do some of the things that the speaker said dogs tried to stay keep out of trouble: sniffing the ground as we walked toward other dogs, for example, avoiding eye contact, and thus possible confrontation.

Once I knew what to look for I saw the signs again and again. Spotting a dog coming toward us, Calgacus would look carefully at the dog, apparently thinking. If we were walking, he'd slow his pace and drop his head to sniff the ground as he walked; sometimes he would stand still and sniff, and other times he would move away, taking us over to bushes to sniff.

I'd had a good teacher and mentor in Ursula when I was learning about clicker training, which meant that I understood the theory of how it works. I knew that Calgacus was more likely to do the things he found rewarding, and I liked the sniffing he was doing far better than the times he reared up and plunged through the air, barking at other dogs, and so I began to click and drop food on the ground for him when I saw him drop his head to sniff. If he headed toward bushes I'd throw food into the bushes, and praise him enthusiastically. We must have been a strange sight on occasion!

I got used to simply watching Calgacus, and noticed more. When

# Canine aggression

approaching other dogs, often, he would turn his head to the side – a quick flick of his dark face, easily missed but unmistakably there. Other times, his gaze would slide to the side so that he wasn't making eye contact with the dog. He sometimes chose to turn away entirely from the oncoming dog, accompanied by sniffing a bush or some grass. I could see how he tended to move slowly if there were other dogs around: his movements considered and careful.

Sometime after I'd attended the talk I was walking Calgacus in a local park, which is bounded on one side by a river and on another by woods. There is a football pitch in the middle, and a ribbon of path runs around the outside of the pitch. Walking the path that day, coming toward us was a little West Highland White Terrier, who was being walked on a lead by his carer. We moved off the path and onto the grass at the side, but I still noticed how the little dog tried to avoid walking toward us. He hung behind his carer until the lead went tight, and he was being pulled forward.

Calgacus had slowed, and so I did, too. We were barely moving as Calgacus briefly turned his head away from the Westie. Calgacus felt tense through the lead, his steps stilted and stiffer than usual. He didn't put his head down and sniff. I was preparing to move further away – since it seemed likely that one or both of the dogs would bark – when the Westie shook himself, white fur spiking out from his body with such force that I expected to see droplets of water fly through the dry, sunny day. Calgacus stopped walking and shook himself as well, and the pair of them did this two or three times each – one stopping and shaking, and then the other. 'Shaking off' is another calming signal that was mentioned in the talk I'd attended, and another of the things that I'd claimed Calgacus never did.

Following their vigorous shaking, the dogs were relaxed, their muscles soft and limbs moving smoothly. The Westie had stopped staring at Calgacus and lowered his tail. Calgacus flicked his head away more often. When we drew level with each other, the Westie barked at Calgacus – a friendly, playful bark, accompanied by a little playful skip. Calgacus looked at the little one and wagged his tail – then we carried on.

Many years after attending that talk I met the speaker again. She didn't remember me, so I approached her and explained where we had met before. Then I apologised, and told her that I hadn't believed her then, but later found that much of what she had said became crucial to helping me understand Calgacus, and take the first steps to resolve his problem.

For a time, I exchanged emails with the organisers of the UK TTouch practitioner trainings, and thought about and continued to practise the TTouch from my book. The small changes in Calgacus were encouraging enough that I decided to again try and find help from a local dog trainer. This time, I looked on the website of an organisation that doesn't care about its members having academic qualifications, but instead insists that they be committed to using kind methods. A subsequent conversation with one of the members was like breathing fresh, mountain air after being stuck in a hot car for hours. Lynne was wonderful, and talked to me about her love of bull breeds, and Buddy, the Bull Terrier cross, with whom she shared her life. I explained about Calgacus' problems with other dogs, and told her what I'd tried to help with this. By the end of the conversation, I had agreed to take Calgacus to Lynne's class the following week so that we could see whether he could cope with it. We ended the conversation with me feeling less alone.

What I remember most about this period in the journey to help Calgacus was beginning to understand that this would not happen in a straight line. I wouldn't find one, single expert who would fix this for us, as the problem was too complex and hard.

# 16 Getting it wrong – again

As well as the guilt I was feeling about Monty's injuries, I made another mistake that haunts me still, and sometimes wakes me in the night, which concerns the harm I caused to Katie. She was four or five months old when Calgacus attacked Monty; a young dog on the edge of adolescence. She had gotten over her nervousness with other dogs, and was enjoying playing with them. A little more prone to anxiety than Calgacus had been as a puppy, Katie was more fearful of things, and screamed when confronted by them, but, generally, she was happy.

I discovered that if I took her out to walk with Calgacus, she was an enormous help, as I could safely let her greet any dogs we met, and allow Calgacus to hide behind me, unbothered. It also allowed me to have a calm conversation with the dog's carer – something that wasn't possible if we needed to use the umbrella to fend off other dogs.

Katie seemed happy with this arrangement, too, seeming to enjoy the chance to meet lots of new dogs.

"He's a bit worried about other dogs," I'd explain, nodding toward Calgacus, "she loves them. Just wants to play."

The other walkers would smile, reassured that everything was okay.

And had Katie been a relaxed and confident adult dog, everything probably would have been okay, but she wasn't. Katie was an inexperienced puppy, in need of guidance herself, and was already showing signs of being anxious. Things might have been okay, too, if Calgacus had not begun to improve; if he'd continued to ignore or chase away all other dogs, Katie could have carried on greeting new dogs, keeping them away from Calgacus, and he could have carried on staying away. But Calgacus began to want to greet his friends again.

On one particular walk, we met a yellow Labrador – who possessed a love of life so often seen in this breed – who Calgacus had known as a puppy. All throughout Calgacus' withdrawal, this dog had never given up trying to get him to play as they used to. That day, Calgacus remained relaxed as his old friend approached: his forehead smooth and free from wrinkles; his ears soft and slightly forward; his tail gently wagging. I encouraged him to step forward and greet the Labrador.

Katie was with us on that walk and, from the corner of my eye, I glimpsed a strawberry blonde streak, as she flew from her position next to me, barged Calgacus aside, and greeted the Labrador. She wagged her tail, play bowed at the Labrador, paddled her feet on the ground, and generally flirted with him.

I smiled and handed Calgacus some sausage because he'd been calm. "How lovely that Katie is so friendly," I thought.

On another occasion, Calgacus and Katie were running in a field with a couple of dog friends. Calgacus glanced to the side – turning his head and dropping his shoulder – inviting one of the others to play. Again came the strawberry blonde streak as Katie raced to get between them to play the game herself. I wanted Calgacus to play with his friends – thought I'd never see him play with another dog again – so called Katie to me and clipped on her lead, feeding her sausages as compensation and giving her my attention.

Watching Calgacus play, I felt my heart fill with happiness, tears welling in my eyes. But poor Katie couldn't cope with this new situation, and screamed and lunged on her lead, unable to listen to me; unable to enjoy the sausages. She was inconsolable. My poor friend: she'd grown up having a job to do – keep other dogs away from Calgacus – a vital, important job. Other dogs coming close to Calgacus was a source of incredible stress for both me and Calgacus, and now she was having to watch another dog be close to him, unable to do anything about it.

This became an ongoing issue that we never resolved. Katie suffered from a series of health problems, some of which made it increasingly difficult for her to think clearly and deal with stressful situations. I did the best I could once I realised my mistake. I began to give Katie and Calgacus lots of walks on their own with me so that Calgacus could practise greeting new dogs and Katie could do so as well, but without the anxiety of feeling that she needed to keep an eye on Calgacus. When I walked them together I chose quiet, open places without the stress of meeting new dogs. As often as possible we combined walks in these quiet places with meeting friends and their dogs, who were familiar. This allowed Katie to begin to learn calmness around the dogs, even if they were playing with Calgacus.

This all helped, and Katie did make some progress but, as her health deteriorated, so too did her ability to deal with stressful situations. In time, she found so many aspects of going out for a walk upsetting that the best thing to do for her and Calgacus was to always give them walks on their own with me so that I could tailor each walk to their individual needs. For Katie I also concentrated on working with a variety of vets to support her health needs.

Katie was the sweetest, most gentle of dogs, who would cuddle up to me on the couch, and was always happy to be hugged. She endured numerous vet visits throughout her life, and saw massage therapists, chiropractors, and a skin specialist. She endured blood taken, her skin scraped, and many other painful procedures, yet always greeted vets happily, and allowed them to do whatever they needed with her.

People in distress were attracted to her. For several months we regularly met a man who'd been sent home from a psychiatric hospital, and was struggling to cope. He would chat to us on our walks, spending time with Katie, and telling me how sweet she was.

One afternoon she and I were sitting on a bench in the town centre, enjoying some time alone together on a sunny afternoon. A man we didn't

# Canine aggression

know appeared from nearby flats. He staggered along the street barefoot, blood pouring from a wound in his head, and when he saw Katie, he pointed at her.

"That is a brilliant dog," he exclaimed, bending down to take her face between his hands and kiss her nose.

Katie wagged her tail, and encouraged him to sit and talk to us, hugging and stroking Katie. He told us that he'd had a fight with his girlfriend, who had thrown a cup at his head and chased him out of the house.

Katie was an amazing dog, for sure.

# 17 Walking a new path

"How long do you think it'll take me to drive to Bath?" I asked my colleague, Emily, who sat next to me at work, "I'm going there tomorrow."

I was travelling there to attend a week's TTouch practitioner training. I didn't imagine that I'd carry on to become a practitioner (which took three years and cost a fair amount), but I wanted to know more about it, and it seemed a week's course would provide that.

"All day," Emily said, "do you know where Bath is?"

"No. I was kind of hoping it was northern England. You know – roughly around where Newcastle is." Was my naive reply.

"It's down near London. The best thing to do is set off early enough in the morning that you can be past Birmingham before three."

Taking Emily's advice, we did set off early. I'd never driven that far on my own, and decided to break the journey by stopping every two hours at a services, which gave me a break from driving, but brought challenges of their own. I'd try and park at the furthest edges of the car parks, getting out and looking around for people approaching. Standing so that I blocked the doorway of the van, I clipped on Calgacus' and Katie's leads, then, holding them in the van, would look around again. I'm amazed that nobody called the police – I must have looked very suspicious, as if planning something illegal.

Each short walk would be spent trying to get to the often small grass areas at service stations while avoiding all the other dogs being exercised. The day wore on until I could focus only on the ribbon of motorway we followed.

Dusk had fallen and the farm was peaceful when we arrived at last. I was trembling, and felt sick from too many hours in the van, too much caffeine, too much sugar, and too many stressful stops with my hard-to-handle dogs. I wandered around, looking for somebody who could tell me where to put up my tent, feeling increasingly jittery. A large, dark dog barked at me from the garden of the house and I looked at him, worrying about the possibility of Calgacus meeting him. Then a woman came out of the house. Slim, pretty, with long hair, she smiled at me.

Calming the dog, she turned to me and said, "I'm Victoria. Can I help you?"

"I hope so. I'm Tracey. I'm here for the practitioner course and I need to put up a tent."

Victoria smiled, pointed me in the direction of where I could erect my tent, and said she'd see me the following day.

With the tent up, my dogs fed, walked and settled down, I went to the large barn next to my tent. There were lights, and, it was clear, people; friendly people.

# Canine aggression

"Hi. How are you?"

"Have you been here before?"

"There is a kettle and tea and coffee over there."

"What kind of dogs do you have?"

The questions and obvious interest in me felt soothing, and, looking back, the first evening was a blur. I can't remember who was in the room; not even how many. The only clear memory I have is of a brief exchange between two people who were reading through a list of participants for the course.

"There are twenty-five dogs on the course this week."

Twenty-five dogs! It hadn't occurred to me that there would be dogs on the course. My understanding was that this was a course for people to learn about TTouch, and I'd assumed it would be classroom based, and that lectures and possibly video would be used to deliver the content. The possibility that there would be lots of practical work for which other students would also take their dogs had not occurred to me. My expectation had been that Calgacus and Katie would stay in the back of the van whilst I was in class.

I swallowed past the sudden lump in my throat. If I had been closer to home, I would have packed up there and then, and left. Instead, I left the light and warmth of the room, returning to my tent, taking Katie in with me and leaving Calgacus to sleep in the back of the van. I lay there with my arms around Katie, grateful for her warmth and frequent licks from her warm, wet tongue, and worried. Images of dogs running at Calgacus flashed through my head – the scenes were dark, as if happening in the moments before a thunderstorm, and they all ended with Calgacus lunging at the other dog and grabbing them, blood dripping onto the ground.

The morning light, usually a time to shake off nightmares, brought more anxiety, and I got up early to walk Calgacus and Katie before anybody else was about. Then I left the farm and drove to a nearby shopping centre, where I sat in the van and ate the breakfast I'd bought there. Sipping coffee, I felt dreadful. I was tired, lonely, and far from home, and wished that I hadn't come. The dogs and I were managing together, and coming on this course had been a bad idea, I decided.

In the end I drove back to the farm, the distance from home the only thing that kept me there. I told myself that if things didn't improve, I could leave after a few days; that I never needed to go back there. I consoled myself with thoughts of escape.

I drove into the car parking area and stopped. A large, Jeep-type vehicle pulled into the space next to me, and a blonde-haired woman got out. She was smartly dressed, wearing good quality boots, and seemed confident and stress-free. I noticed her unhurried movements as she opened the back of her car and spoke to the dogs inside – and felt a stab of pure jealousy. This woman had none of my anxiety. She didn't glance behind her to see who was around; she didn't rush to get her dogs out in case somebody appeared behind her. She spent time sharing a moment of affection with her dogs before putting on their leads. My stomach twisted, the coffee I'd drunk bitter and sharp inside me.

With leads clipped on, the woman ushered her three dogs from her car. Three Spaniels and, to make things worse, I noticed that one of them was an un-neutered male. Bracing myself I got out of the car and forced myself to smile at the woman and wish her good morning.

I left my dogs resting on their beds in the back of my van and went inside.

That week marked the beginning of many shifts in my view as I began learning more about dogs. One of the changes concerned my feelings about equipment that can affects a dog's neck. At the start of that course, Calgacus and Katie wore head collars to which their leads attached to rather than collars. Ursula was a big fan of using head collars, and had showed me how to use them when Calgacus was a strong and lively adolescent.

Head collars are simply a piece of equipment used to lead dogs. The idea is that they work a bit like a head collar does on a horse, giving the person at the other end of the lead control over the animal's head, therefore preventing them from being able to pull hard. Having two large dogs, I found them invaluable, and even more so when Calgacus became aggressive with other dogs. After the attack on Monty, I became dependent on head collars, which in my mind, became firmly associated with safety, and could only feel safe when walking my dogs if they were wearing head collars.

On that first TTouch practitioner training course, whether to use a head collar the way I did was challenged. The discussion was joined by multiple people in the room, and went something like this –

"Two points of connection can make a huge difference to being led for dogs."

"Oh, yes. Remember the last time we were here," one of the participants nudged the person next to her and they both grinned, "I couldn't believe how much difference it made to me to be led with two points of contact."

"So much easier for the dogs."

"Especially if head collars are being used. They have such a strong influence over the dog's head."

"I've noticed how there is lots of tension in the necks of dogs who are just led with a head collar."

Scared, upset, annoyed, and confused by these comments – as well as interested, hopeful, and enthusiastic – I can remember ducking my head and examining a dark, oval-shaped coffee stain on the carpet. I looked out the window at my van – old and battered next to the expensive, four-wheel-drive beside it. "Tracey, what do you think?" somebody asked, "you use head collars with your beautiful dogs; don't you?"

Taking a deep, shaky breath I replied. "I hear what everybody is saying, but I just don't see how I could make things safe for my dogs and for everybody

# Canine aggression

else without the amount of influence that a head collar gives me. Calgacus in particular is so huge – and he really doesn't like other dogs much."

"Seems like a good time to go outside and do some practise with each other." The instructor said at this point.

The practise backed up the comments made in the classroom. When I was set up with leads held to give two points of contact on me, the feeling of being led and knowing what the person leading me wanted me to do was very evident. With just one hand holding the lead, what was required wasn't as clear and smooth.

I knew it made sense to do what was suggested but it simply did not feel safe to do so. My reservations about Calgacus' size and strength remained, and at the next break when I got them out of the van, I used their head collars with the leads attached as usual.

Such was the instructor's skill as a teacher that she didn't ever tell me what to do or become irritated with my repeated challenges and refusals to do what she suggested. Nor did she stop talking about it, either, covering this area more than once. She set up sessions where the teaching assistants helped with the practise of what she talked about, and allowed discussion of the subject to take place among the students on the course.

I could see that what she said was right, but, even so, years passed before I felt confident enough to do as she suggested. Over time, I stopped using a head collar altogether with Katie, instead, attaching her lead to a harness. I never could do this with Calgacus, although he did always wear a double-ended lead with one end attached to a harness. Calgacus and I had travelled a long way from his attack on Monty, learning and growing together, both of us achieving things that few – myself included – would have believed possible. However, I could never completely get over what happened to Monty, and using a head collar when walking Calgacus was part of it.

Now, I don't use head collars or neck collars, partly to try and make lead walks as comfortable as I can for the dogs, and partly for communication. Dogs use their bodies so much when they meet each other. How they hold their ears, what they do with their eyes, the tension in their neck muscles, how they carry their head, are all part of it, and anything that may restrict movement of those vital areas can cause problems between dogs. When dogs need to be on leads, I regard a body harness and as loose a lead as possible as the best compromise to allow maximum communication at these times.

Alice was a fellow student on the TTouch practitioner course. Beautiful – slim with dark red, curly hair that fell to her shoulders, she moved gracefully, and was confident around her dogs. She had three dogs with her, all of whom stayed close to her, and gazed at her with affection – a gaze that she returned. I watched her that week and felt inept. Next to her, I was lumpy, clumsy, and visibly stressed. My dogs were far from well behaved. Calgacus had begun to improve but at that time he was still suspicious of strange dogs, and I spent much of my time avoiding everybody when they had dogs with them. Katie was struggling to adjust to the small changes in Calgacus, and spent her time desperately trying to get any dog she saw to come over and say hello to her.

One evening everybody else had left for the day whilst Alice and I remained, quietly getting on with our end-of-day tasks. Multiple times during the course I noticed how the assistants and instructors congratulated us when we did well, or pointed out when the dogs had achieved something. I saw how encouraged people were – and I felt it myself whenever anybody pointed out to me something that had gone well.

"I really like watching you with your dogs." I said to Alice. "They look so happy to be with you, and you look so happy to be with them. How do you manage to do that? It's something I struggle with."

Alice paused and looked thoughtful. Steam rose from the mug she held and, as we stood together in that moment, I smelled a faint apple and cinnamon scent. I looked down at my own mug, warm in my hand and filled with strong, black coffee.

"I'm not really sure." Alice said. "Maybe it's to do with what we do every day."

She smiled at that point, revealing small, even white teeth, and her eyes became distant.

"I love teaching my dogs new things, and they love learning them. I don't want them to become overweight so I use most of their food every day to do training with them here and there. They don't eat much out of bowls usually because we spend so much of each day doing things together. Maybe that's it? I don't know for sure, though."

I sipped my coffee, appreciating the warmth as I swallowed the rich-tasting liquid. Alice's words turned in my head and I saw a possible answer to Calgacus' problems in her words. I'd become so down about my life that the dogs and I no longer did anything together for fun. Any teaching I did was a functional chore intended only to keep all of us out of trouble on walks, and I wasn't any fun at all on walks, always looking for trouble around every corner – there was no joy in our walks – for any of us.

Alice's life with her dogs sounded so much better, and I wondered if maybe what we needed was more fun ... Standing in the room after Alice had moved on with her evening, I made a promise to myself that each day I'd find ten minutes for each dog, and spend that time exploring with them new tricks and games. Things that had nothing to do with being better behaved when we were outside; things that were just for fun.

That week marked another change in my life. Nobody told me that Calgacus would always dislike other dogs, and I'd seen a different way for teaching and learning to happen. We students were a mixed group – some, like me, on our first time; others halfway through, and a few more finishing up. Qualified practitioners were there acting as teaching assistants and supporting the instructor. The course instructor allowed discussion, ensured that everybody

## Canine aggression

was able to participate, and, importantly, allowed the students to guide her as to when and exactly what she taught each day.

The discussions about head collars and harnesses that I was part of were powerful for me. I wasn't asked or encouraged to change what I was doing; just to think about it differently. Even though nothing about what I did changed during the week, I had the opportunity to appreciate that learning something new – actually taking it on board and acting on it – takes longer for some than others. Learning is contextual, and how quickly or even what exactly anyone will get out of it differs person-to-person. I watched during the course as others embraced the use of body harnesses and double-ended leads, and moved forward quickly with things that I struggled with.

By the end of the week, I felt better than I ever had. I loved the course and wanted more, so made the decision to carry on: to spend three years travelling regularly to Bath, do the required coursework and become a practitioner myself.

It's a decision I have never regretted.

# 18 New friends

Living with a dog who doesn't like their own kind can be isolating. Certainly, it makes friendships that develop through time spent watching dogs play together in the park impossible, and it is difficult to take part in group walks or fun days out. Despite this, I found that helping Calgacus brought new friends into my life – interesting and supportive people, some of whom were fleeting positive influences and some of whom are friends still.

One of the first was Nancy, and we kept in touch even after Calgacus' attack on Monty. Both of us suffered flashbacks about the event, and had feelings of guilt – me for allowing my dog to hurt Monty so badly, and Nancy for not being able to protect Monty. We were able to provide a measure of help and solace to each other as we processed the after effects of what had happened between our dogs.

Once Monty had healed, we decided to meet for a walk, and discussed strategies for helping our dogs feel okay about this, including: walking on opposite sides of the road in parallel if we needed to let the dogs have distance; keeping the walk short if they were struggling; Calgacus wearing his muzzle for safety, and meeting close to Monty's home in the hope he'd feel more secure on home ground. This first meeting was the subject of a great deal of planning. We wanted nothing to go wrong this time.

My hands were shaking as I attached Calgacus' lead to get him out of the van. I tensed, ready for an explosion of barking when he saw Monty – but there was nothing. He stood calmly, looking at Nancy and wagging his tail. For a moment, the two dogs looked at each other, taking in the other's presence, then they turned away from each other, and found bushes nearby on which to pee. It felt to Nancy and I as if they were slightly embarrassed about the drama and violence of their previous encounter, and had decided – in that stiff-upper-lip, British manner – never to mention it again. After all our planning, this was an unexpected outcome but we took it happily and allowed ourselves to be guided by the dogs.

That day we roamed for hours through the city, and the dogs walked next to each other, taking turns to pee on things and snuffling at bushes. After about three hours, Nancy suggested we get a bus back to where we'd started so that we could both go home. Monty was used to travelling by bus and did so often. Calgacus had never been on a bus before, and I wasn't sure exactly how he'd feel about it, but we decided it was worth a go since walking back would take several more hours.

We boarded the bus and stood in the area allocated for pushchairs and

# Canine aggression

bikes. Calgacus sank down onto the floor with a contented sigh, and settled for a snooze. Near us stood a man who looked repeatedly at Calgacus until finally Calgacus met his gaze. The man bent, scratching those soft ears and under Calgacus' chin, whispering to him for a few minutes. When the man straightened, he said, "That is a lovely dog. I'm really quite frightened of dogs but that one is lovely."

"He's quite a big dog to be stroking if you're scared of them," I replied.

I was intrigued. Calgacus had done nothing to invite attention from this man. There were no obvious indications that he would appreciate being touched. He was a huge, frightening-looking dog, and the man hadn't checked with me if it would be okay to touch Calgacus; furthermore, he had a fear of dogs!

"I know," he continued, "I feel drawn to him. I'm having a really stressful time and when we looked at each other, I felt as if he understood. That's silly, isn't it?"

"Not at all." I told him.

I meant it, too. I couldn't see what had drawn the two of them together, but it had been enough to let somebody with a fear of dogs overcome that fear in order to interact with a large dog. Some might dismiss the experience as wishful thinking – a lonely person seeing a connection that didn't really exist – but I'm not so sure. Dogs form strong connections with people, and from my experience are among the most compassionate beings we come into contact with. A study – *Description of the behaviour of domestic dog by experienced and inexperienced people* (http://www.appliedanimalbehaviour.com/article/S0168-1591(09)00195-6/abstract) – carried out by Gabriella Tami (Doctor of Veterinary Science and Master of Etiology) and Anne Gallagher (Veterinary Consultant) suggests that people who are inexperienced with dog behaviour are more able to understand the intentions of dogs than was once thought. I believe that Calgacus and that man shared a moment of real connection, which I feel was helpful to somebody who was going through a hard time.

One of the dogs I met as I went through the TTouch practitioner course impressed me with his calmness. He would lie on his blanket in the training room, not bothering about the other dogs, even if they barked or moved quickly. As my confidence grew with Calgacus, I would look out for calm dogs who we could walk near, and one lunchtime I asked this particular dog's carer if we could walk together.

"This is amazing," she told me, "this is the first time he has ever walked with another dog." She indicated her dog as she spoke.

"Really? But he seems so calm," I said, my stomach clenching at the thought that he might not be the peaceful dog I'd taken him for.

"I know. It's interesting. He has always been a dog who would scream and lunge if he saw another dog. The first time I came here I almost rang my husband and asked him to come and take him home when I realised there would be other dogs here. The person teaching the course convinced me not to."

"That's amazing. I wouldn't have been able to tell."

"I was quite sceptical about TTouch," she continued, "I couldn't believe that it was having such a strong effect. So after that first workshop, I carried on doing it for a while and then stopped. He went back to being a problem not long after I stopped. That's why I'm here. This stuff is making a difference to him."

I smiled. The relief of chatting to somebody who was seeing such positive results with a difficult dog from using TTouch was profound. When I look back, that was a transformative moment – the first time I really *believed* that TTouch was helping Calgacus, and that to continue learning it was a good idea.

There was much to feel positive about at that time. Calgacus was improving, and I was feeling safer. All of the difficulties had not gone away, however, and at other times I still grieved for the change in Calgacus. I found, too, that I felt a surprising amount of envy, when reading about or seeing people obviously enjoying walking and having fun with their dogs. One person on a forum I visited often wrote cheerfully about how her Spaniels liked to charge around on walks, and how much pleasure she got from walking them. Reading her accounts, I would feel a stab of envy. This person had done nothing wrong at all – but she so obviously had what I was grieving the loss of.

My negative feelings about this unknown person were so strong that I deliberately avoided her one weekend when we were both at the same conference. I had gone with a friend of mine – Linda – who I'd met on the same internet forum. Linda was keen to meet another of the group who posted on that forum, and went to meet 'Spaniel woman' during one of the breaks. Grumpily, I stayed away.

I was seeking dogs on whom to practise for my TTouch course, and 'Spaniel woman' posted on the forum to say she was seeking help for a dog from one of the classes at which she instructed, as he was struggling with strangers. He was a large, powerful dog, apparently, and his barking at people in the street had already resulted in a visit from the police. I offered my help, and 'Spaniel woman' – Helen – asked if she could come along to the session.

Helen and I found ourselves talking afterwards, and she offered to meet Calgacus and I with her Spaniels, reassuring me that they ignored other dogs when on walks, preferring to run around following scents.

Helen was all of the things that had irked me in her writing – relaxed, happy, calm – and she and her dogs were perfect for us. Calgacus relaxed once he realised that her dogs weren't going to come near him, and I relaxed, too: chatting and walking with somebody who wasn't at all stressed was *exactly* what I needed. This was all helped by Helen being a lovely person; someone who has a beautiful attitude towards life, always finding interesting things to do with her family, approaching each day with a happiness and enthusiasm that is lovely to experience. That first meeting was the beginning of a great friendship, which has continued for years, and she and her dogs continue to be a supportive and helpful influence in my life, and in the lives of my dogs.

## Canine aggression

The dog from Helen's training club who I went to see improved dramatically with TTouch. His family found that if he became tense on a walk and they moved away from whatever he was bothered by, and stopped to do some of the TTouches I'd shown them, he would often relax. I am no longer in contact with them, but hope that, in some way, I provided help when they, and their dog, needed it.

# 19 Was I able to teach – or not?

I remain ever-grateful to Calgacus for helping me remember that everybody learns differently, that it's okay to be a bit 'alternative,' and take the time needed to learn how to do things. These are all lessons that I've learned multiple times about myself, but it's good to remember that they apply to dogs, too. Calgacus had his own way of doing things, and learning how to help him, and how to build a good relationship with him, meant working out what his way was.

It started with teaching him to bring me things: after all, every dog loves a game of fetch ... don't they?

"Of course you can teach him to retrieve," Linda said, "I'm sure you can."

She stood a distance away from me in the middle of an open field, looking out over green hillsides that seemed to glow in the sunlight.

"He's not interested in toys." I told her.

I was irritated with myself as I said the words, because, by then, I knew the theory –
• Make the toy interesting
• Don't let your dog play with it on their own
• Produce the toy for short, exciting playing sessions with much enthusiasm, and then put it away again until the next time
• Think about the type of toy, and find one appropriate to your dog: one that can be stuffed with food for dogs who really like food, or which has feathers for dogs who really like chasing birds, or which can be attached to a line, and can be moved quickly in an exciting fashion

Alternatively, the advice is to not even try and make the actual toy interesting, but do what is known as 'backchaining' to encourage your dog to retrieve it. Start with the toy in your hand, and click your dog for firstly nosing it, then mouthing it, then holding it, then picking it up and dropping it, then picking it up from a step away and turning before dropping it – until eventually, the toy is retrieved.

I *knew* all this stuff but could not put it into practise, as Calgacus would not fetch. Toys! He wouldn't even look at them – unless another dog wanted them, but even then his interest extended only as far as taking them from the other dog if he could, and then dropping them. He would chase things – rabbits, deer, squirrels were all fun for him to run after – but he would not run after things I threw for him. Calgacus would stare at me, all wrinkled brow and questioning eyes. After a while, he'd turn away with a sigh, as if he found me too silly to bother with. It was frustrating, to say the least.

"You need to make the toys exciting," Linda said, kindly. Her enthusiasm and playful nature are infectious, and her dogs love to play.

# Canine aggression

"I've tried! I've tried everything and he is just not interested. The best I can manage is to get him to mouth a toy I'm holding in my hand. I cannot get him excited about them." I cried.

"Can I try? Is it okay if I try? I have a toy here." Linda is one of my loveliest friends, and her eyes shone with enthusiasm as she looked at me, one hand opening her bag to retrieve a yellow tennis ball so clean that it glowed in the bright daylight. Looking at Calgacus, she crouched down slightly, bouncing from the knees in front of him, and waved the ball enticingly in front of his face.

"Do you waaaannnnntttttt iiiiiitttttttt?" She asked him.

Calgacus looked at her.

"Come and get it," Linda said, her voice high-pitched and excited. She turned away from Calgacus and ran across the field, waving the ball behind her, and squeaking it.

Calgacus watched her run, waited for her to stop, turn, and make eye contact with him, then pointedly looked away.

Linda tried her absolute best, and her absolute best at playing with toys is far, far better than mine. I've watched her numerous times, and learned from her, but I cannot replicate her enthusiasm, the fun she adds to games with toys. But Calgacus was resolute: he did not want to join in, and, eventually, he wouldn't even look at Linda.

Another area where Calgacus' behaviour was far from conventional was when I returned home after being away for any time. According to several well known and respected dog behaviourists, if your dog doesn't get up and greet you when you get home, it can only be because you have no relationship with him.

Arriving home, I would stand in the doorway to the living room, looking at Calgacus lying on the couch to check that his furry sides were moving in and out. Typically, he would lie completely still, not even opening his eyes to look at me in greeting, never mind getting up and running to me.

I would walk to him, crouch down and gently stroke his chin. He would open his eyes, then, wag his tail slowly, and lift one front leg so that I could stroke his belly. After a few moments of this greeting, he would slide off the couch, and we'd go for a walk, or out into the garden.

I have to admit, that at first I found this hard to deal with. I expected that my dog would run to the door and greet me enthusiastically, and worried that he didn't care about me because he didn't do this.

Over time, and partly through the work I was doing that involved learning how to work with Calgacus to deal with his emotional problems, I learned to see this differently. Calgacus had come from a secure home, and was loved by Neil and Susan from the moment he was born. He grew up with an unshakable confidence about his place in the world. He also liked his own space, and demonstrated a strong independent streak from just a few weeks old. His lack of enthusiasm wasn't a rejection of me, but was, instead, a reflection of how Calgacus viewed the world.

I was really into the swing of finding help for Calgacus, going to loads of courses and workshops, and learning lots of wonderful new things, although it wasn't always easy to determine whether or not the help on offer would be useful.

Calgacus and I were once in a group training environment, at a time when he wasn't entirely happy to be in a room full of other dogs, although he could cope most of the time. This particular group was practising 'down-stay,' which meant that all the dogs had to lie down and remain there until their carers went to them, and let them know that it was okay to move. I asked Calgacus to lie down and he looked at me, but otherwise didn't move. I tried again with the same result. Then I asked him to sit. Sitting was acceptable, it seemed, as Calgacus did as I asked.

"Okay, then, sweet pea. Just do a sit-stay. That's fine." I scratched him on the spot that he liked under his chin, and handed him a piece of sausage.

"Calgacus should be lying down," the instructor said.

"I know. But he really doesn't want to. He's a bit worried about the other dogs being around him. I'll just get him to sit instead."

I wasn't trying to be difficult. I understood that Calgacus was required to stay still, and not be running around the class or barking at the other dogs, which would have been distracting or upsetting for them. But, other than that, in my view, the actual position he adopted while staying still wasn't terribly important.

"He should be lying down. He doesn't respect you if he won't do what you say." The instructor insisted.

"He's fine," I said, "he's a bit bothered by the other dogs."

"Look, everybody, Calgacus isn't lying down and Tracey is  letting him get away with it," the instructor told the rest of the class.

I smiled, and shuffled my feet as my cheeks warmed. I felt that Calgacus *was* bothered by the other dogs, but I didn't know that for sure. Was I making *excuses* for the fact that he and I needed to do more work, or perhaps I was allowing him to ignore my requests? This was an ongoing concern for me: the worry that by working *with* Calgacus I was complicit in and encouraging of his anti-social behaviour. I worried that I made *excuses* for him when I shouldn't.

I considered who it was that I was in disagreement with – an instructor I respected, with lots of experience and compassion for the dogs and the people he taught. Also, somebody who – to my knowledge – had no significant experience of working with and helping dogs with behavioural and emotional problems, like those Calgacus had. I considered the consequences of being wrong in my assessment. Maybe Calgacus chose not to lie down because he  couldn't be bothered or wasn't in the mood – but so what, if so? We could do more work, figuring out ways in which I could convince Calgacus that lying down when I asked him to was worthwhile. But if I was to nag Calgacus into doing as I asked, this could well weaken the relationship that we shared.

"Oh" I said, "I don't think he's getting away with anything. He's a bit worried about the other dogs and doesn't want to lie down near them," I told the

# Canine aggression

instructor. Then – trying to inject a little humour – "Anyway, how could I make him lie down if he doesn't want to? He's enormous."

"I'll make him," the instructor said, walking toward us.

That wasn't what I intended. The instructor was supposed to laugh and drop the subject. Thinking fast, I said, "It's just that I'm not sure how Calgacus will take to being made to comply. I'm sure it'll be fine. Of course, if it isn't fine and he objects a lot, I'm happy to take you to hospital to be patched up."

My words stopped the instructor in his tracks.

"Okay, then," he said, "maybe Calgacus can  sit. Right – let's get on with this stay."

In his eleven-and-a-half years of life, Calgacus had never threatened to bite a person, even if they were a vet doing things to him that he really didn't like. But suggesting the possibility that he *might* was the least confrontational way I could think of to put a stop to the instructor's insistence on forcing Calgacus to lie down.

The rest of the class passed peacefully, and in the months that followed, I did work on figuring out ways to make lying down when asked more fun for Calgacus. I found that he particularly liked walks where I would wait until he was a distance from me and then call to him to lie down. Throwing himself onto the ground, he would lie there, looking at me and waiting. After a few seconds I would call him to me, and Calgacus would spring up and run to me as fast as he could. It seemed that lying down was much more exciting and interesting for him if he knew he would get to run fast to me afterwards, and the food and scratches he got when he was with me helped build anticipation.

Another lying down game he seemed to like was surprising, since it meant waiting for good things. With Calgacus standing next to me, I would throw a really tasty piece of food – a whole sausage, chicken wing or a piece of roast beef – ahead of us and let him begin running to it, asking him to lie down before he reached the food. Spinning round to face me, Calgacus would drop to the ground and wait. Sometimes I would go to him and give him different food, or call him to me and give him food. Other times I would encourage him to go and eat the food I'd thrown. Either way, he always got to run to the food I'd originally thrown and eat it. Maybe the anticipation of not knowing quite what to expect after he lay down was the reason why Calgacus loved this game.

As time passed Calgacus became more comfortable in group environments, and would lie down in class,and also at a distance from me, if asked to. All he needed was time to work through his worries until he felt able to relax. To me, affording him this courtesy is the essence of good education, providing time and space for learning to occur. Letting go of the idea that there is only one right way to teach, and deciding that timescales don't matter, are integral to living with the knowledge that – as with Calgacus' greeting of me – it's fine for individuals to do their own thing.

# 20 Letting go

As I worked to help Calgacus feel happy around other dogs once more, I found I had to let go of some ideas and solutions I had used. Knowing that Calgacus was aggressive toward other dogs meant that having him wear a muzzle, and keeping a distance from other dogs – apart from a few select friends who he ignored – made us all feel safe. But in order for him be happy around other dogs, I had to let go of some of the reassurance that went with a muzzle and keeping our distance.

I was dedicating increasing amounts of time, money, and energy to finding a way for Calgacus to be happy – and safe – around his own kind. By this point, I had ended a romantic relationship of ten years, finding myself unable to continue with it and deal with what was going on with Calgacus, and had moved to a smaller, cheaper house. I was totally committed to learning as much as I could about dogs, as well as establishing a support network of people.

I realised that I needed to change how I saw the world, and particularly how I saw Calgacus, in order to notice changes in him, and respond to these positively. I had to believe that his dislike of other dogs was something that could change.

To believe that a dog with a psychological problem may be capable of change sounds like a simple thing to do, but it is, I believe, one of the hardest to achieve, and I know that I am not alone in thinking this because of the many conversations I've had with others who report similar feelings and struggles.

One of the first changes I noticed had nothing to do with Calgacus interacting with other people's dogs, but was about him and Katie, and the fun training we'd been doing in the house after my conversation with Alice during the week in Bath. And we'd begun to interact more with each other on walks. More and more often, on our walks in wide open, bleakly-isolated areas where we could see and avoid other dog walkers, I found that if I whistled them, the pair came running as fast as they could. Part of the fun training we'd been doing had involved toys, and Calgacus had found that, as long as he could throw around the toy, he quite liked them. This enjoyment transferred to walks, where he was also keen to play games of chase and retrieve with toys.

To allow us to play fetch games without them falling out over who should pick up the toy, I had also worked on teaching them that when I threw a toy, whoever ran for it first should be allowed to go and get it. The other dog would happily stay with me and eat sausages, play tug or have a cuddle. To get to this stage, I began by clipping a lead onto the harness of one of the dogs. Then, while I fed that dog, I would say the name of the other, and throw the toy for them

# Canine aggression

to chase. Over time and with practise, the lead became unnecessary, until, eventually, I no longer needed to say the name of the dog who was to chase: the dogs learned to watch each other and decide between them who chased the toy and who stayed with me. More often than not Katie chased and Calgacus stayed with me, but sometimes it was the other way around. I'm sure that their desire to avoid being in conflict with each other helped make this training so successful. I was thrilled with the changes in both of them, and especially in their self-control around the chasing of toys.

One day, we were out on moorland, the mist rising from the ground giving everything a surreal feel. I shouted to the dogs, and threw a tennis ball. I wasn't careful enough with that throw, and the dogs weren't close to each other when the bright yellow orb left my hand and arced up into the damp, grey air. Each dog heard my shout, saw the movement, and both ran after it, each believing they were the first to chase the ball, and expecting the other to be nowhere near it. I stood still, breathing the damp, boggy scent of the sodden ground, unable to think what to do.

As I watched, Katie got to the ball and grabbed it, just a split second before Calgacus arrived at the same spot. He reached for the ball in her mouth, and she turned, snarling, furious that he'd broken the rules and was trying to take the ball that was hers to pick up. The ball fell from her mouth but, by then, Calgacus wasn't interested in picking it up. He charged at her, barging with his shoulder, and barking at her. This escalated to grabbing at the scruffs of each other's neck, both of them making a terrible noise as they each tried to drive the other away from the ball. The rules for playing with the ball had been broken, and both felt that the ball should be theirs.

I watched in dismay, not knowing how to stop them arguing. In the end, I got out my whistle and blew it – one long blast – the sound clear and bright in the morning air. The dogs stopped their squabbling and ran to me as fast as they could! I was utterly delighted, and gave them both liver cake in handfuls. I spent time with each dog, scratching Calgacus at the base of his tail until he wriggled with pleasure, then hugging Katie to me, whispering in her ear that I loved her.

I'm sure that I saw the biggest changes in my ability to observe, and allow Calgacus to learn new skills and get over his problems whilst I learned about TTouch. My first response to TTouch had been a knee jerk refusal that it could be useful. Having softened that opinion, I'd learned to incorporate some TTouch into our lives, though there was still much I struggled with, and groundwork exercises fell firmly into that category.

Groundwork exercises are a movement-based practise whereby dogs are led over low obstacles, around cones, or around a structure called a labyrinth – a pattern laid on the ground made up of piping, rope, pieces of wood, and anything else that can be used for this purpose. Different surfaces – sand, matting, grass, pebbles, wood chip – are often used in TTouch, the intention being to gently allow the dogs to explore what it is like to move their bodies in ways that are not part of their usual habit, so all kinds of equipment can be used. These exercises can be done with two people and one dog to provide an experience of being led by somebody other than their usual carer, without the stress of taking them away from their carer.

The exercises are useful as an aid in observing dogs, too. Does the dog turn more easily to the left than to the right? Do the dog's back feet track behind the front feet, or are they spaced more widely or narrowly? Does the dog's tail wag? Does he sniff the ground? Does he yawn or lie down or roll around?

The groundwork exercises can give dogs a new confidence in their surroundings and their ability to move around. Sometimes, dogs who would normally bark at other dogs can be seen moving calmly around the obstacles with other dogs nearby. For some, the opportunity to move around is vital, as they may feel constrained by being touched, may wish to be active, or simply find having a person close to them overwhelming.

When I first saw groundwork exercises, I wasn't impressed. I saw people and dogs moving around without – it seemed to me – much of a plan or idea of what they were doing. I couldn't see how this would help. When I tried it out for the first time, I felt a bit silly: I didn't know what I was doing, couldn't understand the conversations about it going on around me, and the dog I was working with – Calgacus – was refusing to move.

Luckily for me, I had around me lots of people who felt differently; lots of patient teachers, and I spent time on it. I began to see the differences in some of the dogs, and understand what people said when they talked about groundwork. I could see differences in myself, too: how I positioned my body and handled the lead. It was several years before I stopped feeling awkward and a little ridiculous whenever I spoke about it to people who weren't on the courses with me, but I could see the benefit, even if talking about it wasn't comfortable.

Calgacus, however, remained steadfastly opposed to groundwork. He would sigh when taken to an area where groundwork was to take place, would refuse to move, and simply fix his attention on things going on around him. I decided that his problem was with the effort of having to move, as he preferred to rest or move as he chose, but this was something else I had to question and ultimately let go of. Over time, I discovered that movement filled Calgacus with the deepest joy, making him happier than I'd ever seen him, and providing him with a new way to defuse conflict.

Help from a group of friends enabled me to realise this. I'd become friendly with a few of the people who attended Lynne's class, and some of them were keen to try out heelwork to music – an activity that was fairly new, then, in which people taught their dogs tricks, chose a piece of music and choreographed themselves, their dog, the tricks, and the music into a dance routine.

The star of the class was Laura, and her beautiful, golden-haired dogs, whom she'd trained well as they stood out in class. Laura also had lots of choreography experience, and was happily exploring heelwork away from class,

# Canine aggression

learning new skills and having a wonderful time. Another friend from the class, Carrie, told us that she had plenty of space where she lived to host a training group ... and that was how we got started. We met on Friday nights at Carrie's home to practice together.

We borrowed books and videos to help us get going, and Laura was generous with her time and advice. There was so much to learn and to practice: we worked on teaching our dogs to walk next to us on our left and right sides, and at the same time worked on learning how to position ourselves so that we didn't trip over them when we turned around. The dogs learned to turn in clockwise and anti-clockwise circles, walk backwards, and to weave through our legs as we walked – and many more tricks. No longer was Friday a night to head to the pub, and instead was filled with laughter as we showed each other what we were trying to teach our dogs, and shared our problems about doing so. The room we trained in had dark, well worn carpets on the floor, and tall towers of stackable plastic chairs around the walls. We would unstack a few chairs and each set up a little space for our dogs, treats and belongings.

The dogs were good-humoured about our attempts, especially when Carrie baked before class and produced bags of liver cake or smoked fish treats. The air would be rich with the aroma of fresh baking – Calgacus particularly liked the little banana and peanut butter muffins that Carrie sometimes made.

We had no set format for our evenings, and nobody told us what to teach our dogs; instead, each of us worked on whatever seemed most appropriate for our down animal,and everybody else helped as much as they could.

The movements that the dogs were learning were not habitual for them – or us – which is partly why we found ourselves so clumsy, and the teaching sometimes difficult. The different ways of moving became something similar to the groundwork in TTouch: calm, gentle, non-habitual movements that allowed our dogs to experience their bodies and the world around them in different ways.

Calgacus was bright and enthusiastic; keen to move, his eyes shining. His step would be lighter and his entire body full of joy as he looked at me. And the changes weren't confined to heelwork: he was different, and his sleep habits altered. A month short of his fourth birthday at this point, Calgacus maintained the aloofness he had shown as a puppy, keen to have his own space to sleep in. He was not a dog who wanted to curl up on the bed of his carer ... until one night when I was in bed. Calgacus was lying on the dog bed on the floor, as he preferred, and Katie was in her usual position on the bed by my feet. I had turned out the light, the room filling with that orangey semi-darkness that town residences with street lights outside have.

The mattress moved, shifting under his weight, as Calgacus scrambled up onto the bed. He stood, in the still night-time air, a dark outline limned in orange from the street lights outside, looking at us, seeming unsure of what to do on this unfamiliar surface. Then, deliberately and with great care, he turned in a circle three times before flopping down on the pillow next to my head.

For the next two or three years, Calgacus slept right there every night, his massive body curled in a ball of striped fur ,and his face close enough for me to stroke, although I didn't often do this, remaining mindful of his preference for personal space. At some point, he stopped sleeping on the bed, quietly and without fuss, just as his deciding to sleep there had been.

The experience alone of moving his body in new and novel ways seemed to help Calgacus with other dogs, and one thing in particular seemed useful – bowing – which I spent time over several evenings teaching Calgacus. He'd lower his head and front of his body to the ground, while keeping his back legs straight, stretching out his back. I'd smile to see him in that position, and Calgacus would wag his tail at me.

Our mutual joy at this particular movement wasn't entirely unexpected, as bowing is used by dogs to communicate with each other, with different meanings at different times. Soft-bodied, excited dogs will do it to invite a game (play-bow), stiffer, worried dogs might do it to signal appeasement, and focused, serious dogs might do it to prompt prey to run, and there may yet be other uses. The bow, it seems to me, is to do with initiating movement, but it had never been part of Calgacus' communication with other dogs, even when he loved to play with them. Maybe he wasn't flexible enough when he was young, or just didn't pick it up as a puppy, but I'd never seen him bow, unprompted, to a person or a dog.

One afternoon, as he and I walked together in our local park, I saw a man ahead of us walking with his dog – also a large, male, guarding-breed animal. Calgacus and this dog had never met, but had spent plenty of time glaring at each other from across the park, heads held high and chests puffed out. I took their stiff postures and stares as clear signs that they were irritated to find each other in a place that each considered their own – and I avoided this dog whenever I saw him.

On this particular day the other person had walked far ahead of his dog, and when his dog spotted us, and realised how far away his carer  – who usually prevented him dealing with the interloper in his park – was, he trotted toward us, his stride choppy and determined. I looked around, searching for an escape route. The grass was bright green from several days of rain, and had been cut, so the air was heady with the scent of freshly-mown grass. Calgacus was at the end of his lead, glaring, body stiff and pointing toward the other dog. I called to him.

"Calgacus – come this way."

He ignored me.

I stroked his head and patted his flank.

He ignored me.

The dog was getting closer, his pale amber eyes fixed on Calgacus, who stood taller, the fur on his back bristling as he considered this rival for his territory. Glancing ahead, I could see that the other person still hadn't noticed what was going on.

# Canine aggression

I would happily have turned and left the park to the other dog, but Calgacus wasn't up for that way of dealing with things, and I wasn't confident that I could tow him away fast enough, anyway. The bow movement that he'd been learning popped into my head. Would he do it, though, in the park – a new place for this trick – and whilst feeling annoyed?

I took a breath and held it, feeling shaky. Then I asked Calgacus to bow. He hesitated, looked at me, considered, and then dropped onto his elbows, his massive head close to the ground, so that he was no longer glaring at the oncoming dog. Hoping that this magic would continue a bit longer, I handed him a piece of roast chicken, and asked him to stay in the position.

The other dog stopped advancing when Calgacus bowed, and stood still, looking at us. For a moment, we three stood in a bubble of suspended time and place ... then the spell was broken when the dog glanced over his shoulder at his still-departing – and still-unaware – carer. Turning back to us I could see that his posture had changed: he wasn't as tall; his muscles weren't stiff with tension, and he was no longer glaring from beneath a furrowed brow. He turned and trotted away.

Calgacus stood up and shook himself, and I staggered over to a nearby wall to sit down until my legs stopped shaking. Eventually, the bow became part of Calgacus' everyday language, which he would do when he wanted other dogs to play with him.

I learned my lesson about dismissing groundwork and non-habitual movements, but I fully believed that I'd never change my mind about body wraps, and laughed out loud the first time I read about them in my TTouch book. How could wrapping bandages around dogs in different ways help them? And Calgacus' reaction when I tried this reinforced my view. Once wrapped, he would become still and tense, his eyes showing white at the sides, and his brow furrowing. He'd stand there, striped fur showing around the bandages on the dark green carpet, in a room scented with the biscuity smell of dog feet mixed with the sharper aroma of coffee, and just wait for the wrap to be removed.

In fact, I might have avoided body wraps altogether, but Monty gave me a whole new perspective on them.

To complete my course, I had to find fifteen people who were willing to let me practise my new skills with them and their animals, so that I could write about each in a case study. Nancy – being the wonderfully kind and supportive person that she is – volunteered herself and Monty.

We were both nervous that first time I visited them, remembering that the only other time I'd been to their home had been the day I dropped them off after Calgacus' attack on Monty. I walked slowly up the wide, stone staircase to their door, breathing in the dusty, aged scent of the building, wondering with each step whether Monty would be upset to see me.

Nancy opened the door, Monty held gently in her arms like a small, cream, fluffy package.

"I wasn't sure how he'd feel about you being here," she said, clearly having had the same thought.

Monty's tail wagged furiously, bumping against Nancy's shoulder, and he leaned forward, reaching with his neck to sniff me. Since he seemed to bear no grudge, Nancy put him down on the floor, and he trotted to me, sniffing, wagging, saying hello, before busily bustling off – a little fizz of cream fur and pricked ears.

During the course of my visit, Monty periodically came to me to say hello and let me scratch his ears before moving off again, and, in fact, I found it difficult to do any TTouch at all on him because he was so keen to move. I wondered if a body wrap might allow him to experience moments of rest – I was hoping for calm, relaxed stillness, not the sort of tense and anxious stillness that I saw in Calgacus. With the aid of some cheese, Nancy and I were able to persuade Monty to stay still long enough to put a simple wrap across his chest, over his back, and under his belly.

Once it was tied loosely at his side, Monty lay quietly on the carpet for periods of time, looking relaxed. Nancy and I chatted as we watched him, and I was aware of the buzzing of a fly or bee in the room, which flew across the room – a speck suspended in the air by gossamer wings – passing close to Nancy's reclining dog. Monty lifted his head, and stared hard at the insect for a moment before looking away and relaxing his neck.

"I've never seen him do that before," Nancy said.

"What?" I asked.

"Normally, if a buzzing insect flies into my flat, he runs around barking madly, and pauses only to frantically lick the top of his left leg. His reaction is so sudden and extreme that I have to warn dog-sitters about it in advance when he is being looked after by somebody else."

I nodded, sagely, as though I would have expected the wrap to be calming, whilst, inside, I was feeling a mix of interest and incredulity: could the wrap really have made that difference to Monty?

This enlightening experience with wraps was followed some time after by another. I'd been asked to do a short, informal talk on TTouch for some Spaniel enthusiasts, which took place in an old barn with a dusty, concrete floor; bare wooden beams visible in the walls and roof. The barn smelled musty and damp, and old plastic chairs had been placed there, their numbers swelled by newer, brighter camping chairs brought along by some of the attendees.

The barn was small, and floor space was in short supply once chairs, people, and a few silky-coated, black-and-white or brown-and-white Spaniels were gathered. One of the little Spaniels seemed scared. She sat on her carer's knee, her body spasmed by shudders, and her eyes wide, with the whites showing. I worried, thinking that this wasn't the best place for her.

I decided to start by talking about body wraps. I explained what they are, and how they might be useful, then pulled one from my equipment box and demonstrated the light, stretchy nature of the fabric. I passed it round so that

# Canine aggression

everybody could feel this for themselves. Then, taking time and moving slowly, I talked to the person with the shivering Spaniel on her knee about how she could go about putting the body wrap on her dog.

Inside, I wasn't confident. I felt that the setup was too busy for this little dog, and I didn't think that a body wrap would help. I felt the pull of conflict between the little dog wanting to be anywhere but there, and her carer's desire to learn as much as she could at the talk. I concentrated on the body wrap, talking the woman through how to put it on correctly, pointing out to all concerned how loose it was, resting gently on the dog's skin.

Once we were done, I stepped back, advising the Spaniel's carer to keep an eye on her dog to gauge whether this was helping, letting her and everybody else know that it would be okay to remove the wrap if the dog seemed bothered by it at any point.

We moved on and talked about other aspects of TTouch, but my eye was repeatedly drawn to that scared little dog, who had actually stopped shaking, and wasn't pressing against her carer as hard. A little later, she got down onto the floor, lay down, and went to sleep.

At the end of my demonstration and following discussions, the room seemed to erupt, with people standing up and moving about, talking to each other. Chairs scraped on the hard floor. The dogs were stretching and shaking out, moving around, keen to be outside again. The little Spaniel wearing the body wrap had been asleep, still, but woke to the noise and movement, although she continued to lie, relaxed, on the floor while around her the activity and noise continued.

Time passed and I learned more about TTouch, practised more tricks for our Friday heelwork group, and worked on helping Calgacus discover and more situations where he could turn away from dogs rather than bark at them. Just as I had had to modify my views about groundwork and body wraps, I had to update my view of him as a dog who didn't like other dogs. But, as much as I wanted him to like other dogs again, I resisted this happening to a degree. I'd kept us – and everybody else – safe by having Calgacus on a lead, and sometimes muzzled, when we walked with other dogs, and to consider letting Calgacus off the lead to mix freely with other dogs filled me with dread, and that awful moment when he turned and grabbed Monty played over and over in my mind.

Despite this worry, my new practise of observing Calgacus, and trying to understand him, made me notice new things about him. One night he would not settle down and sleep, repeatedly getting up, out of bed, nudging Katie with his nose and whining at her. Once or twice she indulged him, and got up herself to wrestle with him on the bedroom floor, or allowed him to chase her down into the living room where I could hear them running around after each other. Eventually, though, she stopped wanting to play it was night-time and Katie wanted to sleep.

The hours passed and, far into the night, Calgacus refused to rest, continuing to whine and try and get Katie to play. I lay in bed, my head aching with tiredness, and felt a knot in my stomach. What did it *mean*?

I looked at Calgacus. We locked eyes and he whined – the sound low and plaintive – and then he glanced at the door and backed away from me. The clock said 5:30am but I got up. I knew that Carrie walked early each morning with her two dogs, and, although we had never walked together, Calgacus had known her dogs for a while by then, and saw them twice a week – in class and at our Friday night groups. I hoped she would agree to let Calgacus and I join them on their walk.

Fumbling in the early morning light, and feeling my stress levels rise, I dressed, put Calgacus into my van, and drove to Carrie's house, where I parked outside. Katie stayed at home in bed, and had a walk afterwards. Then, feeling awkward and anxious, I phoned Carrie, explained where I was and why, and asked if it would be okay for Calgacus and I to join them that morning. I promised to keep Calgacus on his lead the whole time if either of us had any concerns, and Carrie was happy to agree.

Setting off, I watched Calgacus carefully for signs that he might behave aggressively toward either of Carrie's dogs. Lily had always been wary of Calgacus, and he had done little to dispel this. Ghillie was young, playful, and male – the sort of dog that Calgacus usually didn't like. As we walked, I saw no signs of stress in Calgacus, who remained calm and relaxed in his body as he walked. I let him off his lead, and he made no aggressive moves toward either of his walking companions. He stopped to sniff, wandered around freely, let the other dogs do what they wanted. It was a wonderful morning.

In the process of writing this book, I've spent much time remembering and thinking about what it was like helping Calgacus, and one of the biggest things was letting go of expectations and beliefs about how the world worked. Even those that kept us safe had to go in the face of some of the changes we both experienced. This wasn't easy to do, and is a process that is ongoing. I still find the need to change my views, to take nothing for granted, and to work to keep my interactions with dogs free from expectation.

# 21 More new friends!

In 2005, Linda told me that she was organising a workshop taught by a dog expert who specialised in helping dogs who don't get along with their own kind, so I went along, and spent a fascinating couple of days listening and watching. Through a mix of lecture, video clips and practical demonstrations, we were shown how breathtakingly subtle dog communication is, and how much dogs can learn from each other. I can still remember driving home from Leeds late on a Sunday evening, the excitement from everything I'd heard keeping me awake for the long hours of driving home along the dark ribbon of road.

The weekend gave me even more incentive to watch Calgacus, and try to regard some of his reactions to other dogs as canine communication. In one of the breaks, I'd spoken to the instructor about Calgacus, and the things I'd observed and learned about him.

"I'd love to be able to help you with this," she said, "but I think you live too far from me. You're in Scotland, right?"

"I am – and I agree. We're too far apart."

"Maybe I know somebody who can help. Do you live close to Edinburgh?"

"I do; it's not far from me at all."

"I'll dig out a number for you before the end of the weekend." She told me.

I returned home with the phone number of a local person who the instructor felt should have the skills and knowledge to help Calgacus and I a little more. In theory, I possessed what was needed to do this myself, but carried with me all the previous experiences where dog experts hadn't helped us, and, in some cases, had made things worse. I was hesitant about again going through the process of meeting somebody to try and determine whether or not they could help, so I simply kept the contact details.

Calgacus was improving, and had begun to play with Katie again. He could go to classes now without terrorising the other dogs, and the walk with Carrie, Lily and Ghillie had gone well. He had learned to be calmer around other dogs, and was more likely to sniff the ground or move away from a dog he didn't want to interact with than lunge out at him. This was much better progress than I'd ever hoped for at the start of our learning together, but I was a bit stuck about how to gain the confidence I needed to let Calgacus mix more freely with other dogs, and also how to help him learn to use his new skills when he wasn't attached to me by a lead, and didn't have my help immediately at hand.

I knew that Lynne had a contact in Edinburgh, too – a man she considered something of a canine genius – so, one night at class, while Calgacus greeted Lynne, covering her in slobber and demanding sausages, I asked for his name, and it was the same as that the instructor in Leeds had given me. A few weeks after that I went to Bath to spend a week learning more about TTouch, and, whilst there, I spoke to one of the assistants – my friend, Pat – about the Leeds weekend and the name – Dave – I'd be given.

"Oh, yes – I know him. He is good." She told me.

I was convinced, then, that this person was worth having a chat with. But still I put off getting in touch. Life with Calgacus was manageable, and I was in conflict. I wanted him to have more freedom with other dogs, but did he really need it, I wondered? He seemed happy enough, and I could not rid myself of scenes of horror whenever I thought about him being with other dogs.

Later in the year, I was attending a conference at which Dave would be present, and I forced myself to speak to him. I still had doubts about the wisdom of progressing with Calgacus, but couldn't ignore the small voice that whispered Calgacus could do more.

Dave and I agreed to meet, and I'm happy to report that I have never been anything other than glad that we did. Having a group of people I trusted to speak to about Dave before I met him made all the difference in deciding that he would be a good influence for Calgacus.

Excitement, joy and hope were the emotions I felt when Dave and I met at a field close to my house. Flat, wide, open and quiet, this was a favourite place for me to walk with Calgacus at that time. Back then, Dave's large dog, Cooper, was young, healthy, and full of fun. His black coat gleamed, and he loved meeting new people. He greeted me by sticking his head out of the car window and licking me. Cooper remained on his lead for the duration of the walk as it was likely that his youthful exuberance would have been too much for Calgacus to deal with.

Also with Dave were Riley – a short, stout brindle crossbreed with wiry fur and a twinkle in his eye, and Cody – a handsome, long-coated dog who was gentle and kind, with lots of experience at meeting new dogs and getting along with them.

We walked around the field, the off-lead dogs running around and Calgacus keeping a distance from them. The long hours of training, the planning, and the many nuggets of sausage and chicken had helped Calgacus realise that he could move away from dogs who were worrying him. He had become quite good at this, and that day was an opportunity for him to practise the tactic with new dogs.

He was also good at following me if I walked away and called him, and so, once or twice when he looked worried about being near to one of the other dogs, Dave would remind me to walk away and call him. This simple stuff – helping him to see that he had options other than to be aggressive – helped enormously in terms of my confidence, and also in terms of Calgacus' ability to be around other dogs.

Dave is fantastic to be around at worrying times, as he has the talent of being able to give other people confidence and a feeling of safety. Even if things

# Canine aggression

are on the verge of going wrong, he can quietly and calmly suggest something that helps. Dave's calm help was key for me in learning to again trust Calgacus with other dogs.

As we walked that afternoon, Cody approached Calgacus, standing nose-to-nose with him and greeting him. I gulped, and could feel my heart beating faster with anxiety. Calgacus could and did ignore other dogs, but to have one so close paying attention to him wasn't something we tended to let happen. Dave calmly told me that he trusted Cody to know what to do when greeting another dog, and that he would not have done so if it was not okay. I believed him, but I could not relax. I couldn't watch them go through their initial greeting and found myself forced to turn away. My hands were shaking, and my breathing rate increased.

I clenched my fists and closed my eyes for a few seconds, breathing in the air, made sweet by the smell of the woods at the outer edge of the field. When I turned back to look, Calgacus was bowing to Cody: elbows on the ground; rump high and tail wagging. They ran together, chasing each other around the field, pausing sometimes to bow or to wait to see if the other still wanted to play. This play was no big deal for Cody– he regularly met and mixed with lots of other dogs – but for Calgacus it was the first time since the progesterone injection that he'd met a new dog, and played with them during that first meeting. More than that, he looked happy and relaxed while playing – just as he had before his troubles – which I thought I'd never see him do again. This was a dream coming true in front of my eyes.

The game went on until Riley became concerned that they were playing too much, and trundled over to them, placing his sturdy body between the two and looking at each of them, asking them to calm down and be more sensible. Calgacus was happy to oblige. That moment was an eye opener for me: Calgacus stopped playing because he could see that it was worrying another dog. He went off to sniff around the field again.

Meeting Dave, and the walk we took that day merit a chapter all to themselves because this was one of those pivotal moments where life changes suddenly and for always. Until then, Calgacus and I had been experiencing gentle, careful change, supported by a range of kind and selfless people. That day was a life-changing experience for Calgacus and I, and marked the beginning of us both becoming more relaxed about meeting new dogs.

# 22 Enjoying play

Calgacus playing with Cody was a turning point in his life. Shortly after this happened, I met with my friend, Sofia, who I'd come to know during the weeks I spent at TTouch practitioner trainings. Sofia had a dog with her – Lex – who I'd never met but had heard lots about, and he greeted me like an old friend, wagging his tail, pushing himself against me as encouragement to stroke his short, plush fur. Lex was a beautiful dog: mostly black, with white chest and shoulders, and rich tan markings on his face. He had ears that stood up and flopped forwards just at the tips. I felt as if I knew him straight away.

"He's not usually as affectionate as that immediately," Sofia commented, watching us, "... usually, he's more reserved with new people."

Maybe, I thought, Lex felt the same way as me: experiencing that odd sensation of meeting somebody new but feeling as if they are an old friend.

We'd planned to walk together with Calgacus and Katie, and my stomach fluttered with nerves as I put on their leads and prepared for the walk. Katie still struggled at times to let other dogs be close to Calgacus, and Calgacus – whilst he was improving – was far from being a happy, playful dog who took meeting new dogs in his stride. Lex wagged his tail when he saw Calgacus and Katie – wide, relaxed sweeps through the air. Katie paused, looked at him, and then mirrored his tail movements, staying calm, her feet firmly on the ground and her breathing slow and even. Calgacus gave Lex short, interested glances at intervals, glancing away and sniffing at the grass and wooden fence posts that bounded the track. We walked together for a long time with the dogs on their leads, careful to ensure they had space from each other. Had they seemed tense, we would have kept it to a lead walk only but there seemed to be some kind of magic in the air that day.

I breathed deeply, filling my lungs with the warm, friendly scent of horses from the fields around us. My jacket rustled as I walked, and I could feel the pressure on my waist from the ever-present bumbag full of treats. We reached an open field; a place where the dogs could be safely off their leads.

"Do you think they'll be okay to have a run together?" Asked Sofia.

"I'll keep Katie on her lead to start with but I think Calgacus will be fine – so long as he has space," I told her.

"That'll be okay. Lex is usually pretty cautious around new dogs and prefers space, too."

My fingers trembled as I unclipped Calgacus lead. I trusted Sofia, and had heard lots about Lex in the time I'd known her. He did sound like a perfect dog to introduce Calgacus – and Katie – to.

Lex was at a distance from Calgacus when he looked up, stood for a

# Canine aggression

second, and then raced across the field to shoulder-barge Calgacus before running off again, clearly inviting a game. At the moment their shoulders met, Sofia and I each took a sudden, sharp in-breath, freezing to the spot, anticipating that the greeting would go down badly with Calgacus. Feeding sausages to Katie to distract her from the tense moment, I held my breath ...

Oblivious to our anxiety, Calgacus bowed to Lex, and then – just as he had with Cody – joined in a happy game of chasing and being chased, Much good-natured barging was evidence of their mutual enjoyment. When the game finished, Katie was still calm and relaxed, so then she, too, was able to run around the field with our new friend.

As he gained in confidence, Calgacus found he could make more friends. Our Friday night heelwork to music group wasn't just about learning new things. We drank coffee and chatted at the end of every group with the dogs who were happy in groups around us. After a time Calgacus could also be with us – at first on his lead but, later, as my confidence in him grew, unrestrained. By that time, Helen had begun attending our Friday night meetings, and one of the dogs she brought with her was her lovely Spaniel, Alby – striking black and white in colour, with hair that is silky soft to the touch. Alby has a beautiful attitude and temperament: happy, playful and enthusiastic about anything that he gets to do. People regularly fall in love with Alby, and want to be around him.

Alby had tried numerous times over the period that Helen and I knew each other to get Calgacus to play with him, but without success. Even so, he maintained a positive attitude, seeming to believe that one day things would change. He was sure that a game could be had, and that all it would take was persistence.

One Friday evening after we'd finished practising, Calgacus was with us as we sipped coffee. At that stage, he was happy to be around the other dogs, though tended not to try and play with any of them. Alby saw a chance. He came over and bowed in front of Calgacus, his eyes full of fun. "Shall we play?" He asked with his unwavering enthusiasm. Calgacus stood up ... and then bowed back.

Both of them were filled with glee as they ran madly around the space, narrowly missing tables and bookcases, and I've seldom seen Calgacus as happy as he was that night. Round and round they charged, stopping sometimes to bow or to shake off and look at each other to check if was okay to carry on. The fun he shared with Alby must have been a huge boost to Calgacus confidence; it certainly was to mine.

And it wasn't just playing with other dogs: something had happened that changed how I thought about Calgacus.

Before our problems began, Calgacus didn't appear to care if other dogs barked at each other, and would remain unperturbed when irritated dogs did this in Ursula's class; also ignoring their carers shouting at the dogs to try and make them stop. This was no longer the case, however, as Calgacus became intensely interested in how other dogs behaved toward each other, and how their carers behaved also. In classes, he would leap up and bark at those dogs who played roughly with each other, or who growled or barked at each other.

Classes for me were exhausting and messy experiences, since I gave Calgacus lots of food should other dogs become excited, trying – though not always successfully – to make his feelings about these situations more positive. I assumed that rising excitement levels in other dogs worried Calgacus to the extent that, should he be able to, he would attack them. I was by now confident that he wouldn't injure them but, for obvious reasons, I didn't want to put this to the test, so continued to keep him on a lead and ply him with food.

However, one Friday evening I was at heel work to music class, where Katie and I were practising our moves in the middle of the floor, while Calgacus – relaxed in the company of close friends – slept at the side of the room, his lead looped under the leg of one of the plastic chairs. On the opposite side of the room, four of the dogs from the group lay close together while their carers chatted. Concentrating on Katie, I didn't see what happened but, suddenly, the four dogs erupted, barking furiously at each other. Calgacus sprang upright and ran – the chair bouncing along behind him – across the room, roaring as he charged the other dogs: his muscles bunched and his eyes fixed on the other dogs. Holding on to Katie, who was upset by the commotion, I felt sick.

The four dogs stopped barking at each other and stood in stunned and motionless silence, as did everyone there, dumbfounded by the speed and noise of his reaction. Calgacus stopped at the edge of the canine group and stood, looking fixedly at each dog, making no attempt at contact of any kind. The others mirrored his stance for a few seconds until a group member, Irene, went to Calgacus, picked up his lead and led him back to me. Returning to our seat, Calgacus sighed loudly and lay down.

Calgacus' actions were food for thought, his behaviour reflecting some of the things I'd learned about at the workshop in Leeds, and during that long ago, half day seminar about calming signals: dogs dislike conflict and will try to keep things peaceful. Calgacus was never barking because he wanted to attack the other dogs, but being on a lead meant he couldn't do anything else. Knowing this revealed much about Calgacus' personality, and gave me a lot to think about.

In various ways – some of them huge, like playing with Cody, and others so small I almost didn't notice them, like the first day he dragged me off into a bush to sniff as a dog passed us – Calgacus progressed to a point I had never thought possible. He now had dog friends who he would play with, and we were, once more, involved in group classes. And I had new friends, too: kind, compassionate people who included me in what they did, despite my large and sometimes problematic dogs. But even more than that, Calgacus seemed to be expressing a desire that the dogs around him should be peaceful and safe.

# 23 Not exactly going to plan

Calgacus, in particular, had become so attentive and obedient, and keen to play at training with me that I felt like I was getting pretty good at this dog malarkey ... and that's when I began to learn that, when it comes to canines, there will always be new things to learn and new challenges to face.

Years earlier, I'd learned a similar lesson from my work as a computer programmer. I've lost count of the number of times I have set off on a path, confident it is the right one, only to find myself a little bit later scratching my head in confusion at what seems to be an insoluble problem. And I'm not alone in this: my colleagues all report similar experiences. Now, after decades in my job, I no longer get frustrated about it. To me, periods of confusion, having to re-think things that turned out to have unexpected results, and not knowing what to try next, are what keeps my job interesting.

Spending time learning about dogs, and figuring out what and how to teach a dog is a similar situation. The dog and the environment change constantly, producing new things to learn and do – and that doesn't stop. A series of events with Calgacus helped me learn this.

I'd become pretty confident about Calgacus' ability to ignore other animals, and to come back to me if he was chasing them. It started with deer, and I spent hours standing or walking slowly with him on his lead, rewarding him for not pulling towards them. I needed him to understand that even if they were exciting, he could not chase them when he was on the lead. I went through many, many bags of treats with Calgacus practising looking at deer or rabbits or squirrels, and then turning and looking at me, or walking only a step or two toward them. This was possible for him, as, without the thrill of chasing something that was running away, Calgacus found that he could manage to maintain control while he was on a lead, and he was able to stop himself from taking off. For a long time I despaired of this self-control ever existing in a situation when he was chasing something. Once engaged in a chase, his focus was entirely on that, and he couldn't look at me or listen to me.

I wasn't hopeful of this changing for various reasons: Calgacus' breed is known to be independent; I was inexperienced with dogs, and Calgacus had chased, caught and eaten a couple of rabbits in his life, which, I was told, would make it impossible for him to learn to stop chasing prey animals.

The breakthrough for me came when I watched a DVD entitled *Really Reliable Recall* from a dog expert called Leslie Nelson. Up until then I'd rewarded Calgacus with small pieces of food for good behaviour – pea-sized pieces of sausage, liver or cheese – but the DVD suggested using many more rewards:

lots of interaction of the type the dogs like; several pieces of food; celebration and admiration for something well done. Turning away from chasing a prey animal would be a big deal for a dog like Calgacus, so he would need more from me than a single piece of liver and a pat.

Having made this change, I noticed a difference in Calgacus. He liked the things I gave him – the food, pats and joy – but, gradually, the act of galloping to me became, in itself, thrilling. As time passed, he and I would go to places with wildlife and practice, until eventually, when he saw a running deer or rabbit, it became an opportunity to run to me and get lots of good stuff.

As he got older, he and I would walk on the moors together, with Calgacus off-lead, and admire the deer, to whom we could stand quite close. I would feed Calgacus and tell him he was amazing. I would ask him to stay before walking for a long distance across the moor, so that I could call him and he could get that thrill of running to me.

Calgacus got so good at this that I became a little complacent on one trip to TTouch practitioner training. Walking the dogs meant taking them up a ribbon of track with short grass at the side, and neat wooden fencing enclosing the horses in the fields on each side. At the top the track opened out to wide fields and, on this particular trip, the field to the right had sheep in it, contained and protected by strips of electric fencing, since that field had no wooden fence. We were there for a week, and for several days, my habit had been to detach Calgacus' lead as we walked into the field on the left. He ignored the sheep completely. Being much more interested in investigating sheep, Katie stayed on her lead. On the last day of the course, I was walking with Sofia who was there without any of her dogs. She had hold of Calgacus' lead, and when we reached the point where I would normally let him run free, the field was empty. I told her that she could let him off his lead.

"The sheep?" Sofia queried.

"Ah – it's fine," I said, "Calgacus ignores them." I replied, my tone slightly smug.

Sofia unclipped his lead – and Calgacus whipped round away from us and ran toward the sheep, pushing his way between the strands of electric wire, ignoring the shocks he must have received, and joyfully continuing toward the sheep. I could see his intent: he wanted to play, and was running toward the sheep in hope of a game. Of course, the sheep didn't want to play with a dog, so were gathering together and running. They looked frightened and I was worried. Calgacus could harm the sheep, who are injured easily and stressed by being chased, plus, any farmer would be well within his rights to shoot a dog chasing their sheep.

I pulled out my whistle and blew on it.

Calgacus stopped running after the sheep, and turned to run toward me. His feet flew over the ground and, when he got to the electric fence, he barged through it with huge enthusiasm, showing no more concern for it than he had

# Canine aggression

when he'd gone in with the sheep. I was so happy that he ran back to me, but this event taught me a lesson about dogs and sheep. Now, no matter how much my dogs seem to ignore sheep, I keep them on their leads if there are any nearby.

Proud that Calgacus had came back to me when I called him, it wasn't long, however, before I began to notice problems resulting from his enthusiasm for doing so.

After a trip to Bath, where Calgacus had enjoyed a week of being admired and petted by several lovely people, coming home and finding that most people ignored him was, I think, a bit depressing, and he hit on a method to stop this happening. I noticed his new strategy when he barked at a lady who was crossing the road away from him. While barking at people isn't ideal, I wasn't worried about this as I knew he was attention-seeking, and I was pretty sure that it wouldn't produce the desired result. Who would ever give attention to a huge dog that was barking at them, I reasoned? Quite a lot of people, it seemed, amongst them a roofer who was working on the house opposite me, who had descended his ladder, and was walking down the garden path as Calgacus and I left home. Calgacus barked, and jumped up and down, and the roofer came straight over to us and patted Calgacus. With his hand caressing Calgacus' ears, he asked "Does he bite?"

"No. He won't bite you." I told him.

I was frustrated that Calgacus had apparently learnt that barking at people was a good way to get attention from them. Probably around eighty per cent of the people Calgacus barked at either came over and patted him or stopped, looked at him, smiled, and had a conversation with me.

The problem escalated when Calgacus discovered that he could use this strategy to make me call him when off the lead. If we saw a cyclist or a person he liked the look of in the park, he would run toward them, barking, knowing I was guaranteed to call him and, when he came to me, would give him loads of food and attention: ergo Calgacus received attention from new people, plus he got to play his favourite game of running to me, and getting lots of fantastic, tasty treats into the bargain. Even better (for him) but worse (for me), some of those he barked at would then come over and talk to him – sometimes they'd even have a biscuit for him!

Once Calgacus had figured this out, it became a lifelong source of fun for him, and a worry for me. We managed to keep it to a minimum by carefully choosing when and where to walk off-lead, and by practising when in busy areas with Calgacus on a lead being allowed to say hello to people when he hadn't barked at them.

I knew that calling him and rewarding him for coming to me was also making running and barking at people more fun for him, but I could not let a dog – especially one of his size - run at strangers and bark at them, as some would have been terrified. I also knew that allowing people to talk to him after he'd barked at them made the barking more likely, but if I'd refused to do so, those people may have gone away thinking that Calgacus was dangerous, which could have got us into serious legal trouble.

Although this situation was annoying, I learned something of great value. Change is always possible and learning is always happening – for dogs and for people. More than that, I learned that to have an explorative, enthusiastic dog is to have a companion who is interesting, informative and – at times – challenging. A life shared with Calgacus gave me the understanding that many dogs have firmly held opinions, and a thirst for learning about how the world works.

# 24 Confidence grows

With Calgacus becoming more playful with my friend's dogs, and enjoying meeting Dave's dogs, I found myself feeling more open to new experiences, and being around more people with dogs. I'd joined the much-loved forum for Spaniel enthusiasts that Helen ran, and, when I logged on one day and saw a post suggesting a group camping holiday, I commented that I would love to go along. I remember taking my fingers off the smoothness of the keyboard, sitting back, my hands resting on rough denim jeans, and feeling the butterflies in my stomach. People and dogs: who would be there, and what if it all went wrong, and Calgacus or Katie couldn't cope with the other dogs?

So much time has passed since that holiday that I don't remember much about it. I know I kept Calgacus on his lead, and we didn't mix much with the other dogs. I know that Katie had to be walked on her own to prevent her becoming over-excited. I remember watching a sea of Spaniels running around, tails wagging and pink tongues lolling. I still have a photo of Calgacus taking part in a competition for who could fetch a tennis ball the quickest. In the picture, he is leaping over piled up hay bales, soft ears flapping and mouth slightly open with the effort – a broad, muscular tank of a dog who made the photographer shake in case he didn't stop in time. He wasn't the fastest. Even if he hadn't become over-excited by all the cheering and dropped the ball, he wouldn't have been the fastest ... but he did have fun.

Toward the end of our holiday, he made one new friend: a Spaniel called Mitch who was fast, young and playful. His rich red-and-white fur streamed out from his body as he ran. This dog was the embodiment of the sort of dog that Calgacus had found most difficult, but when Calgacus met Mitch, he wagged his tail and encouraged Mitch to come closer. They had a little play, and afterwards I was so happy I felt as if I was floating. I remember that brief meeting because of the progress it represented for Calgacus, but had no idea then how much it had meant to him.

Two years passed before Calgacus saw Mitch again, on another Spaniel forum holiday, where there were around 30 Spaniels. I was more relaxed about Calgacus by then, and he spent a lovely time hanging around with several of them. There was plenty of space, too, for Katie, although she was calmer on that holiday. On the last day of the holiday I was packing up, and had taken Calgacus for a quick walk. Suddenly, he stood still and stared, ears forward, head tilted to the side. Preoccupied with how much I had to do to get everything into the van, and the long drive ahead, I didn't look at what held his attention. Calgacus planted his feet, leaned further toward what he was looking at and wagged his tail. His breathing quickened and he whimpered a couple of times – quietly and urgently. I looked then and saw that he was staring at Mitch. One Spaniel in a sea of Spaniels, but Calgacus picked him out, and clearly wanted to say hello. We hadn't been camped near Mitch's family, and Calgacus and I must have been walking at different times to them, which meant that the two dogs hadn't come across each other previously that weekend.

I took Calgacus to Mitch, calling out to Mitch's people that Calgacus wanted to say hello. The joy that the dogs felt at seeing each other again was clear for all to see, as they sniffed each other, wagged their tails, and stood close together. That each remembered the other, and they were delighted to be together again was very obvious.

What I'd regarded as a fleeting greeting wasn't how Calgacus and Mitch felt about it. They'd established a bond on their first meeting, which was still there two years later, which made me realise how important it is to pay close attention to the social encounters of dogs, since even the briefest can be incredibly significant.

Calgacus had become so relaxed and happy around Spaniels that I was keen for him to walk with them, and so, when one of my friends got a new dog – a Spaniel – I told her "She's lovely. We should take her for a walk with Calgacus."

I smiled as I said this, remembering the many times I'd almost drown in anxiety if I saw a Spaniel in the park close to my home.

"I'm not sure. She's scared of big dogs," was her uncertain reply.

"I think it would be okay. I can keep Calgacus on his lead if it helps."

So, early one morning, with the mist clearing, and the dew drying on the grass, we met in a field to walk together. The little Spaniel ran round, following smells, tail wagging as she took it all in. Calgacus watched, his eyes following and his head drooping, as though watching her made him tired. From that point, it seemed that she considered Calgacus safe, and whenever she saw him, she would run to say hello. Calgacus, mindful that she might be frightened by him, would carefully turn his back as she approached, allowing her to sniff his bottom before turning to greet her.

More and more often I was looking for dogs who Calgacus could play with, and it took a little adjusting to. I'd spent so many hours wishing that he would want to play with other dogs, and still carried with me the thought that staying safe meant avoiding other dogs. I again discovered new insight and support from among my friends. I'd met Pat on my TTouch course: she had been a TTouch practitioner and skilled dog trainer for many years, and has one of the most beautiful smiles I've ever seen. Pat works hard to help people become the best they can be, and she became a good friend; somebody who would listen to me talking about what was happening for me and my dogs.

On one course, Pat had a new dog – a five-month-old puppy called Kira – who is an incredibly sweet and caring animal. I've had the great privilege of spending lots of time with her, and have witnessed her regularly being caring

# Canine aggression

toward the other dogs in her family, as well as with Pat and every other human in the family.

Back then, Kira was young and lively, and had been living with a group of large dogs. She loved other dogs and she loved playing with them. Her youth, confidence around large, adult dogs, and exuberance made me concerned about her meeting Calgacus. I worried that she might push too hard for a game, and that he might become upset with her, so I kept him away from her right up until the end of a week of TTouch training, when he seemed to really want to play with another dog. He became livelier, really looking at the dogs he saw, wagging his tail and apparently interested in them.

I asked Pat: "I know this might seem odd, and please say no if you don't think it'll be okay, but I wondered if you would let Kira and Calgacus have a play together?"

"That would be great," Pat replied, "Kira would love it."

We took the dogs to an enclosed field, Kira bounding enthusiastically ahead and Calgacus walking more slowly, and took off their leads. Kira raced to Calgacus and tried to get him to play. Calgacus sniffed a bush. Kira rushed at him again, leaping and jumping around him. Calgacus moved and sniffed another bush. My anxiety mounted, and I wondered if I'd done the right thing in approaching Pat.

When it got too much I said, "I'm sorry. I thought he wanted to play. I must have been wrong. I'll take him away."

"Wait." Pat said. "I think he's waiting for her to calm down. Just give them time."

So we waited as the sniffing and frolicking went on. After a few minutes Kira sniffed a bush, too (she must have eventually wondered what Calgacus found so interesting about them). She sniffed next to him and he turned to her, wagged his tail, and play-bowed to her.

The game was on, and they chased each other around the field. Calgacus rolled onto his back, encouraging Kira to hold him down. They had a whale of a time, and I remember their play session to this day as a time when listening to the wisdom of Pat's words helped me to learn something new about my dog – that he could be patient with younger dogs, and that he was capable of encouraging them to be calm without using force.

Calgacus had become more comfortable in general with being around other dogs. On one TTouch course, I talked about our heelwork to music practise; how beneficial it had been for Calgacus, and the similarities I saw between this and groundwork.

"Could you show us?" I was asked.

An ancient tape recorder was produced for the music, and we did our demonstration in a small space between stables, the training room, and a barn, surrounded on all sides by watching people and dogs. A pair of elegant Lurchers snoozed in the sunshine, watching us from beneath almost-closed eyelids. Calgacus glowed in the sunshine, his eyes gleaming: the audience was there just to watch *him*. He fixed me with his intelligent gaze and did everything I asked of him perfectly. My beautiful, stripy friend wagged his tail and he pranced as he moved in harmony with me, ignoring everything else going on around him as we moved together in a bubble of concentration. Calgacus and I were never spectacular at heelwork to music; it was something we enjoyed, and I'd noticed was beneficial to him, but so many people and dogs are better at it. His unique story – and his breed – were the reasons for our rapt audience that day. Not many Bull Mastiffs dance, and not many have the opportunity to unlearn aggression toward their own kind.

Not all his interactions with other dogs were positive, however, and one problem became more pressing once Calgacus was interacting more with other dogs. He was hugely attractive to un-neutered male dogs, who would approach him, their eyes gleaming, apparently in love with my handsome hound, and sometimes so much so that if their carer tried to take them away from Calgacus, they would growl.

I remember the first time this happened. The other dog was a young Rottweiler who'd fallen for Calgacus in such a big way that he ignored Katie's attempts to play, and refused to leave us. I was feeding Katie and talking to her to keep her calm about this dog being close to Calgacus, when the young dog's carer came to collect him. When the dog didn't respond to his cues, the man reached down and took hold of the dog's collar, which caused the Rottweiler go let out a low, warning growl.

All of us tensed. It's one thing to have another dog be overly-affectionate to yours, but another to have that dog begin growling.

"Stay," I said to my dogs, feeding them sausages and liver bread in the hope that they would remain calm.

"No," The man said as he continued to pull on his dog's collar, ignoring the growls.

This scenario continued for what felt like hours but must actually have been seconds until there was space between us. My legs shook as I staggered away with my dogs.

This stuff is tricky to deal with, and wasn't at all pleasant for Calgacus. Some of his friends could be a bit over-friendly with him sometimes, which he was okay with. He'd push them away and growl a bit if they went too far, but, generally, he dealt with their attraction for him in a good-humoured way. With strange dogs, though, it really wasn't all that fine, and I felt for him. It was too easy for the situation to become tense when the dogs didn't know each other.

Left to his own devices, Calgacus would have dealt with this absolutely fine. He would have moved, pushed the dog away, growled himself, maybe shown his teeth. The problem was that this happened regularly in places where Calgacus was on a lead, and the other person had to pull away their dog. I was concerned that, in those situations, the other person could be bitten, which I didn't want to happen.

# Canine aggression

Over time we made a deal. If it was safe to do so, Calgacus was let off-lead and he could handle the situation himself. He was good at this, and could sometimes persuade his new acquaintances to play with him, instead of sniffing and licking him. If it wasn't safe for him to be off-lead, he would stand completely still and ignore the other dog. I'd give him lots of food for doing this, and we'd wait until the other dog's carer retrieved them. In this way, they might be bitten – one lady was, by her little Terrier when he objected to being dragged away – but not by Calgacus.

When I began to notice all of these small changes, all the moments when things went well, or we found a new playmate for Calgacus, or a tricky situation went well, I regarded Calgacus in a different light. I'd always loved him, of course, but to see my beloved dog being happy and relaxed, and to feel that I could once more trust his reactions, was one of the most special times in my life.

Visit Hubble and Hattie on the web: www.hubbleandhattie.com
hubbleandhattie.blogspot.co.uk
Details of all books • Special offers • Newsletter • New book news

108

# 25 **Stress and bother**

Dave and I had been friends for a while when he rang me one day. His voice on the phone was tense, and I immediately took my hands from the keyboard I worked at, and turned to look out of the office window, watching cars speed by on the motorway, small enough at that distance to seem unreal.

"Cooper is seriously ill. He needs a blood transfusion today. Can you bring one of your dogs to my vet now?"

"Yes. Tell me where to come. I'm at work but will leave right now and pick up one of the dogs on my way."

My hand shook slightly as I pressed the button to end the call, then I sat for a moment while, around me, people talked, typed, and got on with their day. I was at work – a place where stress comes from system faults resulting in a temporary loss of service, or a need to fix some incorrect data. At work, I know how to deal with these events and, although they can be stressful, they tend not to make me tremble or go still. The difference was that Dave's phone call wasn't about work. Medical emergencies aren't usual, and his call caused me sudden stress. To anybody watching, it may have seemed as if I was hesitating before going to meet Dave and Cooper, but this wasn't what I was choosing to do. Responses to sudden and unexpected stress tend not to be entirely within the control of the individual.

After that moment, I picked up my car keys and handbag and went to my boss. "I need to leave for a few hours. My friend's dog has to have a blood transfusion as quickly as possible, and I'm taking one of my dogs to their vet to donate. I'll make up the time later."

My boss nodded and smiled and wished me well, telling me to drive carefully, showing the flexibility that comes with an understanding of lives – and emergencies – outside of work.

Calgacus and I rushed to the vet to meet Dave and Cooper. I have given blood numerous times, so expected that the whole thing would take about fifteen minutes, and that I would wait at the vet and take Calgacus home soon. It turned out I was wrong. The nurse explained that they would have to first test his blood to ensure that he was healthy enough to be a donor, and that they would then take the blood and afterwards let him rest. She told me that they expected the entire thing to take a couple of hours, and said they would phone when done. She assured me they would ring if Calgacus became distressed after having blood taken.

"Give him a couple of biscuits and somewhere soft to lie down, and he'll be absolutely fine." I told her.

# Canine aggression

With nothing else to do, Dave and I left Calgacus and Cooper at the vet and went to have lunch while we waited. An hour or so passed before the vet phoned to say I could pick up Calgacus.

"How was he?" I asked when I got there.

"Really calm." They said. "He let us hold him, and stayed sufficiently still that he has no bruising at all from where the needle went in. You'll see on the shaved part of his neck. He ate some biscuits, had a drink, and lay down for a sleep. He was great. I'll go and get him."

Calgacus led the nurse into the waiting room, spotted a couple sitting close to me ,and took her to visit them in case they had biscuits they could be persuaded to part with. Then he looked over and noticed me.

His ears dropped, smoothing his forehead, and his tail wagged. He came to me and positioned himself so that I would stroke him under the chin. I whispered to him that he was amazing, then picked up his lead to leave. With that crisis averted, and subsequent good treatment, Cooper recovered to enjoy many more happy years with Dave.

Stress is not only experienced by people: dogs feel it, too. What fascinates me is that people and dogs show very similar responses to being in situations they find difficult, and are commonly referred to as the four Fs: fight, flight, freeze, and fool around.

Fight and flight are the responses that are probably the most familiar and often seen. The dog who barks, growls, or bares his teeth is usually responding to a stressful situation by using the fight response – as is the person who sounds their horn or gestures angrily at another driver. The dogs who run – or try to run – away are in flight mode, as is anyone who has found themselves walking away from or out of a stressful situation.

Freeze can be more difficult to spot, but this is what happened to me in the moments after Dave's phone call. The same can happen with dogs faced with a stressful event – they become very still – and is often missed by people or ignored because the significance isn't known. The dog isn't *doing* anything, after all.

Then there is fool around; the stress response of the class clown. Dogs in fool around mode seem happy: they madly cavort about their people, grabbing leads, sometimes, or jumping up or rolling. They can seem happily engaged in play, wilfully ignoring their carer. The dog or person engaged in fool around mode isn't coping well, and feels under pressure, but their behaviour is quite likely to annoy those around them, which then only increases their stress levels.

The lines are blurred between these stress responses. For example, a dog who has frozen can sometimes explode into aggressive behaviour, particularly if their stillness has not been noticed, or mistaken for relaxation. Fooling around might morph into stillness or running away. Each dog and each situation is different and, for this reason, we need to know and understand what the signs are, in order to know how to respond appropriately.

# $26$ **Back to school**

The adventure that Calgacus, Katie and I were on enveloped me, and I wanted to learn more, experience more, do more. It felt as if I was only scratching the surface of the subject matter, so, in the autumn of 2005, the dogs and I embarked on a three-year foundation degree in canine behaviour and training. (I include the dogs in this because the course had a strong practical side to it, and students were expected to have a dog of their own who they could train.)

Bishop Burton College in Yorkshire was where the course was, and somewhere I found impossible to reach in under five hours. Degree courses in animal behaviour were available closer to home, of course, as well as a degree course in dog training, but I chose this particular one partly because some of my friends were also going to do it, partly because the course had been written and was managed by somebody who wrote on a forum I frequented, and who impressed me with her commitment to a kind approach to the dogs she came into contact with, and partly because of the practical side to it.

The course was part-time, and mostly done by distance learning, with four trips to the college each year. In the event, the three years stretched into five when I decided to spend a further two years doing a top-up course to eventually gain a BSc with honours in Canine Behaviour and Training. In this time I acquired new friends and experiences, and help and support through some challenging times with Katie.

During a week of teaching at the college we looked at some of the activities people spend time doing with their dogs. I can't remember now the detail of everything that we had the opportunity to try, but one particular activity sticks in my mind: fly ball.

We stood outside in the college outdoor training area, shivering in the autumn air. The space was enclosed by a waist-high white fence, and the flooring comprised small pieces of rubber that the dogs loved rooting through to find dropped treats. Jumps and tunnels and high walkovers surrounded us as we stood with our dogs.

"For anybody who doesn't know what fly ball is, it's basically a retrieve game. The dogs run down a channel over three small jumps. At the end, they push on a machine which causes a tennis ball to be released. They grab the ball, bring it back to their handler, and then the next dog goes. Fastest team wins," explained the instructor.

"Right," he continued, "I'll show you how to start teaching this. Tracey, bring Calgacus down to the box."

He indicated a black contraption that stood at the end of a row of three small jumps.

# Canine aggression

"Will he fetch a tennis ball?" The instructor asked.

"Yes. He does quite like bringing balls to me," I was proud to tell him. I'd been inspired by Linda's confidence that I could get Calgacus to retrieve, and he and I spent many hours working on this until he was happy to do it.

"Okay. Well, first, I'll trigger the box so that Calgacus can hear the noise, and we can be sure he isn't scared of it."

I remembered how Calgacus had reacted to Linda's encouragement to join in, and wasn't sure this was a good idea.

"He won't be scared of the noise." I told the instructor. "It'll probably be best to just put in the ball and move on, because he's not always that tolerant of baby steps."

"No," the instructor held up his hand, "we need to be careful about this sort of thing."

"Okay," I said. Who was I to argue? He was the expert, "you're probably right."

"Calgacus," he said to get my dog's attention, then slowly moved to press his hand on a ledge on the device. There was a loud click as the mechanism engaged. Calgacus looked at me.

"Really?" He seemed to be saying with his eyes.

"Good. He didn't seem bothered. We'll move on to the ball now. We'll let him figure out that if he puts his foot on the lever and presses, the ball will fly out."

Waving the ball close to Calgacus' nose, in a high-pitched, excited voice he told him: "Loooook. A baallleeee. Do you want the balleeeeee?" After a few seconds of excited talk and ball waving, the ball was placed in its little holder in the machine.

"Use whatever word you usually do to tell him to retrieve."

"He knows to touch things with his paw: do you want me to ask him to do that?"

"No. Just use whatever word you usually do to tell him to retrieve. He'll have to touch the box to get the ball."

"Calgacus – get it." I said.

Calgacus looked at me, irritation apparent in his calm, brown eyes. Next, he looked at the instructor, holding his gaze for a few seconds, then stretched his neck over the fly ball box. Being careful not to touch the trigger, he put his mouth over the tennis ball and lifted it from the holder. He turned and dropped the ball into my hand and waited for his treat. His disdain hung in the air like a physical weight.

"Ah, yes. I didn't think of that. He's big enough to just reach the ball, isn't he?"

"So he is." I replied. "Thanks for the lesson. I can see what he'd need to do, so if we do decide to do more fly ball, I'll know how to teach it."

When it came to choosing what we would work on for our dog training assessment that year, I didn't even consider fly ball. Instead, since I had a few friends with gundog breeds, I decided it would be interesting to look at teaching him some of the work they would do.

Calgacus had to search for a toy in a field, for which we used a gundog dummy. These come in different shapes and sizes, but the one I chose was a traditional design – a log-shape made from plastic-wrapped sawdust covered in green canvas. The most important part for my purpose was the eyelet at one end with a toggle attached, as explained shortly. I had added rabbit scent to the dummy – not real rabbit but a synthetic scent sprayed from a bottle.

When the scent of the dummy reached Calgacus, and he was sure he knew where it was, he had to stand still and look towards it so that I knew he'd found it and where it was. When I gave him the cue, he had to run toward where the scent was. As he ran, I would make the dummy move via the low-tech medium of tugging on the toggle at the end: the dummy 'scurrying' across the grass away from Calgacus was the signal that he should lie down without verbal cueing from me.

He then had to wait there in the grass, watching the dummy's progress as it continued to move away from him. Once I'd moved it a good distance, I would drop the string, let the dummy lie still, and then, after a couple of seconds, let Calgacus know that he could go and retrieve it.

He came to love this game, which was probably helped by the fact that we would practise the finding, looking at, running toward and then lying down parts with the seagulls who would sit on the football pitch at our local park early in the morning. The gulls were routinely there, clustered at the end of the pitch like a packet of mints dropped on a green carpet. Calgacus would find them for me and look at them, holding still and quivering a little with excitement.

Given the cue, he would run toward the gulls with enormous enthusiasm, making them whirr up into the air in a mass of white, flapping wings. Dropping to the ground, he watched them go. Of course, he couldn't retrieve the gulls, who wheeled and spun in the sky above us, squawking their disapproval at being disturbed, so I'd throw a toy for him to fetch instead.

Test day came around and we gathered at college, milling nervously in the car park, unable to focus much on anything other than our anxiety about the test.

"I know you all want to see what each other is doing, but make sure you ask before watching somebody's test." The course manager was keen that nobody was put under the additional pressure of an audience unless they were happy about it. As we waited, some of the others approached me.

"Would it be okay if we watched you and Calgacus? We're all fascinated to see what he's going to do."

"I'd be delighted." I replied.

I was thrilled to be asked this. Calgacus and I had worked hard on training for the test, and I was pleased that some of my friends would be there to watch how it went. More than that, I felt it would be helpful for Calgacus.

# Canine aggression

One of the test conditions was that the dog could not be given food, or have games with toys as a reward for the duration of the test. Being unable to have a bumbag of treats with which to regularly reward Calgacus throughout the test did make things harder, but I knew that Calgacus liked an audience, and the feeling that he was the centre of attention that this gave him.

A small group of us walked to the field where the test would be held, Calgacus trotting alongside, his posture alert; tail swinging gently side-to-side.

I asked my friends if they would be okay to stand at the end of the field where I would hide the dummy for Calgacus to find. One of the assessors told me subsequently that my friends had been worried about this, in case Calgacus become distracted, and went to them to say hello and ask for food. I'd wondered about this, too, but, in the end, thought it more likely he would prefer to have them watch and admire him.

Calgacus was held by the examiner and distracted with pieces of sausage so that he didn't see where I hid the dummy. When I got back to him, he was watching me, his eyes sparkling and his ears tilted toward me. I took off his lead and sent him to look for the dummy. He ran around the field, a dark shape moving over the bright green of the grass, and then he positioned himself directly in front of our audience and stood still.

"On you go," I told him and Calgacus ran toward the dummy.

This was my cue to drag it through the grass, watching it bounce and slide over the uneven ground. The moment it began to move, Calgacus threw himself onto the ground, glancing at the audience as he did as if to ensure they could all see just how amazing he was being. He lay there, muscles tense, ears forward, eyes fixed on the moving dummy while I dragged it to me, picked it up, and threw it further away from him.

"Go get it," I told him.

Calgacus leapt up and ran flat out across the field, grabbed the dummy and charged back to me, his tail wagging and a look of utter happiness on his face. We got a good mark in that test – 78 per cent – and an A grade. But better for me than this was watching my dog running to me with the dummy clamped in his mouth, and his tail wagging.

When the second year of the course began, I had to decide what to choose for the dog training test at the end of that year. This would be a test of our advanced dog training skills, with a much higher standard expected than in the previous year, and there was no question in my mind this time: I was going to do heelwork to music with Katie.

Katie was fit and healthy and doing brilliantly. In June 2006 – just a few months before the start of the second year – she and I had performed a routine at the first kennel club-registered, heelwork to music competition held in Scotland. Katie came second in the beginner level of the sport known as 'starters,' winning a rosette and a beautiful write-up from the judges –
"This team also had the WOW factor, it must be a first for a Bull Mastiff in the

HTM ring and a first for winning out of starters. The routine was called 'Girls just want to have fun,' and they certainly did, along with the rest of the audience."

Building on Katie's success, I planned to extend the length of the routine and add in some complexity to make it suitable for the college test, which required a higher level than the starters competition.

I can still remember how desperately sad I felt when things didn't work out because, in December 2006, Katie became unwell with a disparate mix of symptoms that included painful, bleeding feet, barking at people and other dogs sometimes, and more regular instances of sudden fear of everyday objects. While she and I went through the process of having tests and seeing several specialists, I felt that it would be unfair to put her under the pressure of preparing for the test. Even though I planned to do all I could to keep things fun and light-hearted, I knew that I might become stressed if I felt things weren't going well and I would certainly be anxious on the day of the test. I wanted to save Katie any worry from picking up on my stress levels.

This meant that Calgacus would need to do the test with me which was a bit of a concern. He loved doing heelwork to music training, and showing off to our friends, but during competitions he had never joined in, and instead simply stood and stared at me while the music played. The college test might seem sufficiently competition-like that he would do the same in that, standing still and looking at me with those wise, brown eyes.

I made a huge effort to avoid this eventuality. We joined new groups for some variety, and, where it was okay to do so, I let Calgacus play with the dogs in the classes to try and make it more fun. I spent hours working out our routine and then breaking down each move into segments, which I then turned into a training plan so that I could gauge if we were on track. I followed the plan and made notes after each training session. And I worried. I had no idea if my efforts would make any difference.

Fears about Katie's health infused that year, causing greyness to seep into our lives, making everything hard-going. Working so closely to the training plan made me anxious we weren't making as much progress as we needed to, and, as the months ticked by, this got worse. Much of the joy I'd found in our dancing disappeared for me – and, I suspect, for Calgacus, too, since I must have been stiff, intense, and no fun to be around much of the time.

The day of the test arrived that summer. I'd slept badly, and was tired and nervous. We went into the training barn at college and stood on the gritty, concrete floor while I tried hard to remember the routine, hoping I'd know when to weave around the pillars, and when I should sit so that Calgacus could lean in and press his nose to my ear as if whispering a secret.

With us in the room were several observers and a video camera. We were the centre of attention, and Calgacus moved around the room happily greeting everybody, wagging his tail. While he did that, I hid food behind some bags on the table at the end of the room. I could give him nothing during the test but wanted

# Canine aggression

to be sure I had food to hand when we were done so that we could go straight outside and reward him.

And, after all the preparation and worrying, that's all I remember of the test! Calgacus did join in, and we got a good grade, so it must have gone well, but the test is a blank to me. What I do remember most is overwhelming relief that it was over with!

One of the lessons that college taught me was that, if allowed to, dogs have a voice, and often form strong opinions. Calgacus had an intolerance of boredom that made one of the modules at college more difficult. In second year, alongside advanced dog training, we worked through a module aimed at teaching us how to organise and teach dog classes, and the test involved taking two short, ten-minute classes. One class had to cover something simple, such as teaching a dog to sit, and the other had to cover a more complicated exercise. Whenever we were at college during the second year, we would practise for this test, forming into classes for each other, with our dogs being taught each others' exercises. Calgacus got on fine with the more complicated classes, and particularly seemed to enjoy an excellent class on tracking that one of the students conducted.

The simple classes were not like this, however, and the boredom he felt when he realised that he was going to be taught how to sit *again* emanated from him in waves. He would sigh, look away, and refuse to acknowledge anybody who spoke to him. I dreaded those practise sessions. One particular day that involved practising for this test, Calgacus managed two classes, and then became disruptive, barking whenever he heard people outside the barn which – given that this was a busy college – happened a lot. He looked around to see what else he could do ...

He looked at the little dog next to us – a dog he liked. Even so, he was upset enough about what was happening that his affection for her didn't matter in that moment. He looked at her and then he woofed at her – a loud, deep bark that sent her under the chair to hide. That was enough. I took him out of the room and put him in the van. His reaction to boring classes was heard and responded to. In behavioural terms, his actions may have been called an extinction burst and, yes, maybe if I'd ignored him, he would have given up, endured the boredom without further complaint, and been less embarrassing. But this behaviour wasn't something it was okay to ignore. I borrowed another student's dog for the next couple of classes, and then Calgacus, having had a rest, was willing to join in again for another couple of classes – so long as I maintained a close-to-constant supply of sausages to make up for the annoyance.

I don't want to pretend that hearing and responding to the opinions of our dogs makes them easy companions – it doesn't – but what it does do is allow for a richer, deeper relationship, and it certainly opens the door to possibilities that wouldn't otherwise exist. I'm sure that had I chosen fly ball as the subject for our first year college test, Calgacus and I would not have gotten an A grade in that test. I listened to his opinion of fly ball, and I listened, too, to what he liked to do, and found something that got us a good grade, and, more than that, became a source of fun for both of us for many years to come. Most things worth having are not easily come by: living with an opinionated, critical dog falls into this category for me.

Visit Hubble and Hattie on the web: www.hubbleandhattie.com
hubbleandhattie.blogspot.co.uk
Details of all books • Special offers • Newsletter • New book news

# $27$ **Best friends**

After the attack on Monty three years previously, and Calgacus' ongoing problems with dogs he didn't know, I had become resigned to not being able to bring a third dog into our home. Then, one day, Helen spoke to me about a dog she was fostering – a little white, un-neutered male dog called Bertie.

Bertie had become disabled, and his owner had felt unable to care for him any longer; despite this, he was one of the happiest dogs I've ever met. Bertie was small enough that it wasn't difficult for him to get around, even with limited mobility. He loved everybody he met and other dogs, too – even dogs much larger than he were greeted with enthusiastic sniffs and a wagging tail.

Helen was keen to find him a new forever home, which turned out to be in Bath. Since I was travelling there to attend TTouch practitioner training, and would be close to the new home, Helen asked if I would take him with me. I agreed: I was keen to help. The only thing was that it meant Bertie spending a night in my home so that we could set off early the next morning. He had met Calgacus and Katie on a camping trip, so wasn't entirely unknown to them, but was still the first dog I had brought into the house since the day that Monty was attacked.

I didn't want to confine Bertie to a room on his own for the night, so took him into the living room with his little travelling crate, and first of all let Katie meet him. Calgacus rested in another room and Katie was on a lead for safety, given the size difference between her and Bertie. Katie was joyful, delighted to have a tiny little male dog around, and lay down to be calmly greeted by Bertie, wagging his tail.

Then, feeling a little nervous, I took Katie out of the room and brought in Calgacus on his lead. He lowered his massive head and sniffed at the tiny, white dog. Bertie wagged his tail. All of Bertie would have easily fitted into Calgacus' mouth if he chose, and for a second I froze, worried about what might happen. Calgacus, having sniffed Bertie and said hello, calmly turned away, got on the couch, circled three times, then lay down and went to sleep. Bertie, tired by all the excitement, went to sleep in his bed in the little travel cage. We had a peaceful evening with Calgacus sleeping and Katie flirting with tiny Bertie, and then an easy drive to take Bertie to his new home.

The situation with Bertie had been so calm that I began to wonder whether it might actually be possible for another dog to come and live with us. Calgacus seemed again to be enjoying the company of other dogs, and there had been good news for Katie, too.

Throughout the previous year, she and I had been supported by my excellent vet who had worked hard to find an answer to the strange symptoms which had prevented Katie from being my training partner for the second year college test. She'd been examined and tested, with impressive attention to detail, and found to be mostly very healthy – other than a bacterial infection in her feet, which was initially treated with antibiotics. When she wasn't on antibiotics, however, Katie developed abscesses between her toes, which would swell, burst, and then heal. Her behaviour wasn't what it usually was, either.

To help deal with what might be happening with Katie, my vet referred us to a fantastic specialist dermatology vet. He was diligent in questioning me about Katie, and about the changes I'd seen in her. He did his own tests and found that she had furunculosis (infected hair follicles) in her feet. He noted that, physically, there wasn't much to see, but was very interested when I told him about the changes in her mood and behaviour. I discussed with him her sudden strange fears – which included rooftops. She would often look up at them, put her tail between her legs, and try to run away. She'd become so excited at seeing squirrels and birds that she'd bark at them madly. She suddenly lacked focus when she and I were training, and was finding it hard to learn new things. She had begun to bark often at dogs in classes – even dogs she knew well and liked. I'd stopped all group training with her as a result of these sudden issues.

The vet told me that furunculosis in dogs can be caused by allergies, and said that, as well as giving her more antibiotics, he wanted to start her on an exclusion diet to see what difference that might make. So, for six weeks, Katie ate nothing but pork and sweet potato. I was so keen to get the diet right for her that I taught her to wear a muzzle for that time so that she couldn't scavenge any food on her walks.

After two weeks on her new diet, Katie's behaviour had improved significantly. She could remain calm around squirrels in the park, and she stopped barking at her dog friends. Roofs became things to ignore once more. Her feet had healed completely, and there had been no sign of an abscess forming. She was like her old self again.

When the six weeks were up, we went back to visit our dermatologist vet to deliver the good news. The next step was to add foods back into her diet an item at a time to see what she reacted to, but here we ran into an issue. Katie's physical symptoms were abscesses on her feet – which would typically take weeks to form. This meant that adding back in foods and waiting to see what she would react to could take a very long time, and could have caused problems in itself, since a diet consisting almost entirely of pork and sweet potato is not a balanced one.

I had gone to the vet with a list of behaviour changes I'd seen in Katie, so suggested that it would be clear to me from her behaviour whether she was eating something she was allergic to. After some detailed discussion about Katie's behaviour and what I would look out for, the vet agreed that the best approach was to add each food item and then look for behaviour changes.

When the process was finished, it was clear that Katie couldn't eat

# Canine aggression

wheat, rice, chicken, beef, lamb or fish. I found her a dog food that was based on pork and potato – both of which she could eat – and started feeding her on it. With the diet change, her feet healed and her mood improved: the old Katie was back!

Time passed without me doing anything but think about a third dog. I'd decided that I would like a small companion breed – probably a Pug – a cute, friendly little dog who looked like a miniature Bull Mastiff. Having worked through the problems that Calgacus had had, and those which Katie had developed, I wanted an easy dog; one who would be friendly with everybody, and not require lots of exercise, so that they could accompany Calgacus on walks as he got older. I spoke to people with Pugs when I visited Crufts, and emailed others, seeking a reputable breeder.

"What about a rescue dog?" My friends would sometimes ask me.

I had an answer to this. I didn't feel confident about bringing an adult dog into our setup. As good as Calgacus and Katie were, they were big dogs, and could be intimidating. Katie needed to be around dogs who would not be defensive with her because she could not back away from conflict. I felt it would be easier on all of us to have a puppy.

"You get puppies in rescues," was the obvious response to this.

It's true. You do. Lots of them. Even cute puppies find themselves homeless and being cared for by dedicated staff in rescue centres. I had an answer to this, too: it was important to know where my puppy had come from; who the parents were, and that the puppy's early life had been relaxed and pleasant. Puppies, whether born in rescues or handed into them, were more of an unknown quantity in these respects.

Any dog we took in would live in a small house in the middle of a town, with primary schools at each end of the street; meeting other dogs, people, and lots of children would be a daily occurrence. Taking on a puppy whose mum may have been worried by people, other dogs, or simply bemused and upset to find herself in a kennel or foster home with her babies was not the start in life I wanted for my puppy. I was wary of things going wrong, and felt that I would be unable to cope with another difficult dog. So I knew what I wanted: a Pug, or possibly another small dog. A male puppy. I knew exactly what I was after.

Then I went to a talk at one of my local Dogs Trust re-homing centres, after which there was a tour around the centre. I have volunteered in rescues a lot over the years, spending many hours with rescue dogs, and whilst I often feel a bond with them, I don't feel inclined to bring them home, so I didn't expect the tour to affect how I felt about my next dog. I could not have been more wrong.

I looked at the dogs there, and what struck me was how okay most of them seemed: normal dogs who had fallen on hard times, and needed the chance of a new home. Wasn't that what I wanted? A normal dog? A friend for Calgacus and Katie? A companion for me, who would, hopefully, not need the medical care and attention that Katie did, and who would – again, hopefully – not have the social conscience that Calgacus had? A little dog for us all?

I went home that day and rang a friend who worked for Dogs Trust. I described to her the sort of puppy I was looking for – a male, one who wouldn't grow up to be a large dog, one whose mum loved people, and who didn't bark at other dogs. I told her that I expected to have to wait for the right puppy to come along, and that was okay.

She described a litter that had recently been re-homed – a bunch of little Collie crosses. Their mum was incredibly sociable with all the humans she met, while ignoring other dogs, and all the puppies in the litter had been males. We discussed that litter, and agreed that the next time they had puppies similar to these, I would come and meet them.

The next day my friend from Dogs Trust rang me back and told me that one of the puppies had been returned after just a few days in his new home. He was living, at that time, in her house, and was getting along fine with her dog. Would I like to meet him? Of course I would!

I needed some support with this as I hadn't expected a puppy to become available so quickly. I knew just who to ask. Carrie, who I walked with, and who hosted our heelwork to music group, has the most amazing observation skills, but, not only that, had shown me incredible kindness. I knew that she would help me spot whether there was anything to be concerned about with this puppy, and, if there wasn't, she would help give me the confidence to take him home.

A chubby, fluffy, dark brown puppy was carried into the room in which we waited, and he stretched his neck to sniff each of us, before being placed on the ground. Once there, he happily moved around the room, sniffing chair legs and exploring the scents on our legs and bags, trotting from one to the other, wobbling and stumbling as he did. He was just six-and-a-half weeks old, still learning to control his legs. He seemed happy, confident, and able to move freely – exactly the sort of puppy I was looking for.

A couple of days later, I went to the rescue centre to collect the little puppy, going from there to Carrie's where Calgacus and Katie were to meet them for the first time. The pup stood in what must have been a strange place, and looked at the bookcases, cafe area, tables, chairs, and all of the people who had come to meet him. He wagged his tail and greeted everybody.

I went out to the van to get Calgacus. As we walked side-by-side to the door, Calgacus moved loosely, his tail wagging slowly. This was a place he associated with fun and games, and I'm sure he anticipated good things happening. We went in and the puppy waddled over, little tail wagging, and stretched up to sniff a delighted Calgacus. Keeping the first meeting short and happy, I took Calgacus back to rest in the van and brought Katie into the space. She wagged her tail, and padded her front feet with excitement when she saw some of her favourite people. When the puppy appeared to say hello to her, she became more animated, sharing sniffs with him and laying down so that he could easily say hello to her. Again, this first meeting was intentionally a brief one, and I returned Katie to the van.

# Canine aggression

The puppy and I stayed to share coffee and conversation with our friends: by the end of the afternoon he had been christened Cuillin. I saw an adventurous spirit in him and a wildness in his eyes, so naming him after one of the wildest parts of Scotland seemed the right thing to do.

This homecoming was nothing like Katie's had been. I had learned from that, and made efforts to encourage peace between the dogs from the start. I had cleared away any toys that were particularly exciting, and instead left out lots of fairly boring plastic chew toys and rope toys that Calgacus and Katie weren't particularly interested in. I made sure that there were far more toys than there were dogs in an attempt to show my two that there was plenty of everything to go around, and they didn't need to feel anxious if they saw one of the others with a toy.

At that time, my dogs were not fed much from bowls, and we didn't use them every day, but I made a point of using bowls every day for a while to allow me to work on encouraging each to stay at their own bowl. To help with this, I put extra, especially-tasty food in the bowls as the dogs were eating – particularly for the faster eaters. It helped that Calgacus and Katie already knew to remain at their own bowls so I really only had to be vigilant with Cuillin. I also spent time feeding them by hand and teaching them that if food dropped on the floor, looking at me would get them more food, teaching them that looking at me was more beneficial than trying to grab the dropped piece of food.

I believe that the living environment matters enormously, and I did everything I could to set up our home in a way that helped the dogs be kind, respectful and gentle with each other rather than competitive and argumentative.

My experiences with Bull Mastiffs have shown me that they are dogs who enjoy rough, full contact play: the sort of play behaviour that might not be found in the normal repertoire of dogs. Certainly, it seemed that this wasn't something that Cuillin understood, so Calgacus took it upon himself to teach his new little friend how games should be played.

During the first couple of days of Cuillin being with us, Calgacus lay on the ground and rolled over onto his back, extending his head to encourage Cuillin to grab his neck in play and hold him down. Cuillin bounded around on his chubby little legs, trying to get Calgacus to get up again.

After a few iterations of this game, Calgacus sighed and climbed to his feet. He reached down and put his massive jaws around Cuillin's furry little body, gently rolled Cuillin onto his back, held him there for a second or two then let go. He got down on the ground again, rolled onto his back and extended his head. He repeated this a few times until Cuillin 'got it,' and when Calgacus lay on his back, he clambered on top of Calgacus, putting his paws around Calgacus' neck and holding him in place.

In the years that followed I'd watch them play, and would see Cuillin leap at Calgacus' shoulder and push him. Calgacus would respond by rolling onto the ground. I wondered if he ever regretted teaching that trick to Cuillin, but could never see any sign of it. Even if he'd been sleeping on the sofa and had been nagged by Cuillin to get up and play, he always seemed to be smiling. For the rest of his life, Calgacus' attitude toward Cuillin seemed to be that Cuillin could do whatever he wanted, and Calgacus would be fine with it.

Visit Hubble and Hattie on the web: www.hubbleandhattie.com
hubbleandhattie.blogspot.co.uk
Details of all books • Special offers • Newsletter • New book news

# 28 A model of disobedience

To suggest entering into a dialogue with dogs, discovering what it is they want, cutting them some slack when they behave badly, and taking their feelings into account flies in the face of many common beliefs about how dogs and people should relate to each other. This chapter suggests there are times when disobedience in our dogs can be a good thing, as I believe it can be for people, too.

Stanley Milgram's experiments highlighted, in a frightening way, where total obedience to authority can lead. Even discounting this, I know that unquestioning obedience is a tricky thing to deal with. I have long experience of working as a computer programmer, and the thing about computers is that they always do exactly what they are told to do. Every. Single. Time.

Their blind obedience is part of what makes computer programming such a challenging field to work in. I can say with confidence that any time a computer goes wrong and refuses to give money at cash machines, or calculates wages incorrectly, or an on-line purchase doesn't go through properly, it is because a perfectly obedient computer followed an order given by a human who had not properly thought it through.

Similar things happen all the time with dogs. There are a couple of times that stand out for me when I noticed my own failure to think things through, and was thankful for a dog who felt able to disobey my request.

Three years had passed since the first time I went to Bath to begin training as a TTouch Practitioner, and I was preparing for my last six-day course. I hoped that, by the end of the course, I would be a fully qualified practitioner. Helen got in touch with me about a little Spaniel, Ralph, who she had helped home. The people who had taken him felt unable to cope with him: he was unsettled, spending his days running in circles and barking continuously. He had been homed outside of Scotland, so was a long way from those who could help him, but I would be passing close to where he was on the long journey to Bath. I offered to pick him up, take him with me on the course, and then bring him back with me at the end of the week. I had space in my van, and didn't think that Calgacus and Katie would mind having a little dog share with them. (This particular story took place before Cuillin moved in so I don't mention him here.)

Ralph was no bother at all while we were away. He settled nicely in the van, curling up and sleeping. At night, he shared my bed and remained settled all night, waking in the morning to stretch his little front legs up around my neck, hugging me close. As the course progressed, he became more able to cope with all the people and dogs around him. He stopped spinning in circles and barking, and seemed less stressed. Some of the others on the course suggested that he might like to play with another dog. I wasn't sure: for all that Ralph seemed to be calmer and more able to deal with life, I didn't get the feeling that he wanted to play with another dog. After a couple of offers of playful young dogs, I said that I would take him out with Calgacus. By then I trusted that Calgacus would give Ralph space if he wanted it – which I suspected he would.

I took both of them to an enclosed field to see if they wanted to play. Calgacus looked at Ralph, lowered his head in inquiry and then dropped into a bow. Ralph moved behind my legs, making no effort to join in. Calgacus tried again. Making his limbs loose and floppy, he loped slowly away from Ralph, glancing back over his shoulder, encouraging Ralph to chase him. Ralph stood next to my legs like a small, four-footed shadow. Clearly, he did not want to play, and preferred to avoid engaging with another dog.

Having tried to start a game and seeing that Ralph didn't want to play, Calgacus went off, sniffing bushes, chewing on grass, and generally amusing himself in the field. I waited to see if things would change and, after a period of time, concluded that the dogs were content to share space with each other, but that Ralph very obviously didn't want to play. I whistled for Calgacus, who lifted his head, looked at me ... and then ignored me.

I was used to Calgacus running to me as fast as he could when I whistled him – I was used to it being such a source of fun for him that it caused us problems. I whistled again – and again, Calgacus didn't move. He sighed as he made eye contact with me. Muttering about disobedient dogs, I stomped across to Calgacus to put on his lead. When I moved, Ralph came with me, staying close to my legs. I noticed him there and the penny dropped. I sighed sharply in annoyance – at myself.

Of course, Calgacus wouldn't run to me! I had a dog clinging closely to me who Calgacus had tried to interact with, and who had made it very clear he wouldn't welcome any kind of interaction. I'd briefly forgotten about the social situation we were in, and had asked Calgacus to do something that wouldn't have been okay with Ralph. Calgacus, though, hadn't forgotten this, and had assessed the situation as one where I was mistaken, so had refused my request.

When I got to him, I gave him a sausage; scratched him under the chin in the spot that he liked. "Good dog. You're such a good dog," I told him as I clipped on his lead.

One more instance of thoughtful disobedience sticks in my mind. On a bright evening, Calgacus and I turned up at a local class so that we could be tested for the Kennel Club's good citizen silver test. This wasn't our usual class, so neither of us knew any of the other dogs there. As we pulled into the car park, I could see the class participants practising for the test. That night everybody would demonstrate that their dog could walk nicely on the lead, get in and out of the car in a controlled manner, come away from distractions to go back to their carer, and remain in one position for two minutes.

We joined in with the practice, and almost immediately Calgacus spotted

# Canine aggression

her: a pretty little black-and-white Collie who was looking at the other dogs. Her mouth was tightly closed, the hair over her shoulders stood up just a little, and her tail curled high over her back. Most of her attention was on a young and bouncy Labrador who had smooth, soft-looking, butter-coloured hair, and who was enthusiastically bouncing to the end of her lead to greet the other dogs. Calgacus looked at the two dogs, his brow a little furrowed and his neck stiff. I called his name and he turned to me, so I got a treat out of my ever-full bag and gave it to him. We joined in, doing some practice with the others. I spoke regularly to Calgacus during our practise, and asked him if he could turn his attention away from the pretty Collie and her focus on the Labrador He received lots of treats, praise and scratches over the course of our session for doing as I asked.

I felt pretty positive about this test. We'd prepared for all of the exercises; Calgacus was responsive to me, and, although he was concerned about the Collie and the Labrador, none of the dogs there would be a huge problem to him. Plus, I had plenty of treats to keep his attention on me between exercises. (A feature of these tests is that the dogs cannot be rewarded whilst they are being tested, but usually can be in the down-time while the other dogs are being tested.)

The examiner arrived and we went into the hall to start.

"Treat bags over here on the stage," she told us, "you can have them back at the end of the night."

I'm pretty sure I went pale. The whole test was likely to go on for an hour-and-a-half, and I was not in the habit of asking Calgacus to behave well in classes for that length of time without food. Scratches, pats, applauding him, and all the other things I did that Calgacus liked would only go so far. I was sure that pretty soon he'd look for food, which might cause difficulties.

Confidence drained from me with each step as we filed into the hall. The first exercise we did was the stay: two minutes of remaining in one position. We found a suitable spot on the floor and I asked Calgacus to lie down. He looked at me pointedly, and then he looked at the Collie, who was staring at the Labrador, and then back at me. He sat and made it clear that lying was simply not an option. He needed to keep an eye on those dogs.

"Stay," I said, giving him a last stroke on the soft spot between his eyes.

I stepped away the required five paces and stood still. Calgacus looked around the room, monitoring things, checking that none of the dogs was going to be unpleasant to each other. A minute into the stay, all was peaceful in the hall. Other than Calgacus, all the dogs were lying quietly, and it was clear that there wasn't going to be any trouble. Calgacus sighed – taking a massive breath that expanded his rib cage and was then forced noisily out of his nose – then lay down, content that he could stop monitoring for the time being.

The next exercise was one called 'Come Away From Distractions.' For this part of the test, everybody stood with their dog in a group in the middle of the hall. One by one, each person unclipped their dog's lead, walked away, then called their dog to them, before putting the lead back on and rejoining the group.

In the group, the young Collie was closer to the bouncy Labrador. Some white showed in her eyes, and she intently watched the Labrador's movements, staring fixedly at the other dog. Calgacus was uneasy. To him, this situation looked like one that could become unpleasant – an opinion I shared.

"I know, sweet pea," I whispered to him, "try to ignore them. Here, let me scratch you and see if it helps."

He turned away from the dogs, still unhappy, glancing over his shoulder and appearing tense. However much I agreed with him, I wasn't running the class, and Calgacus really couldn't do what he wanted, which was bark loudly at the two younger dogs and send them to opposite corners of the room. I worked to keep his attention on me, and he tried to do what I wanted, while ensuring that the tension between the Collie and the Labrador didn't build into confrontation.

Eventually, Calgacus had an idea. With another sigh, he slid down onto the floor, positioning himself so that his hindquarters lay between the two worrisome dogs. He looked at each dog in turn as he lay down – long, considered looks that seemed to say "This is all my space. Don't you dare move into it."

The younger dogs understood and did as he asked. The Labrador stopped pulling toward the Collie and the Collie stopped glaring at the Labrador.

"He's a lazy one." The guy next to me said.

"You'll struggle to get that one to move." Somebody else added.

I smiled at them. "I know. He's so lazy."

When it was our turn to do the exercise, I unclipped Calgacus' lead and walked away. When I called him, he got up and sauntered over to me, his back swaying slowly side-to-side like a lion's.

The other exercises went smoothly, and, although neither Calgacus nor I put a foot wrong, we didn't pass, of course, as changing position during the 'stay' is an instant fail. People commiserated with me, and I nodded and smiled, promising the examiner I would try to train my dog to be more obedient.

But inside I was beaming. Calgacus had shown an enormous amount of concern for the well-being of the other dogs in the class, as well as a desire to listen to me and do as I asked, and had demonstrated a huge degree of free will and creativity in figuring out a way to combine those two objectives. When I retrieved my treat bag, I gave Calgacus everything in it for being so amazing. Knowing that Calgacus would choose to look out for the safety of dogs he'd never met before – even if it meant disobeying a request from me – mattered so much more than passing a test.

I floated home, feeling elated, and the feeling lasted for days. We did the test again a few months later with a different group of dogs ... and passed it without incident.

# 29 The science bit

"It's pretentious nonsense," I said – or something along those lines – "why on earth do people need to know all the scientific terms for things? The dogs don't care, and being able to spout a load of jargon doesn't help anybody."

Linda and I were planning to go to a conference together, and were chatting about whether or not it was important to understand the technical meaning of words and phrases commonly used in behavioural science. I thought not, and thought that for a long time – until I started my degree, and had to learn it all for the essays we had to write.

I've changed my view since then, and now feel that knowing and keeping up to date with the theory is essential, especially when it comes to reading books and papers written by scientists and academics. It is very useful to be able to understand theory well; to talk about it with others, and search for ways that it might apply in life.

After three years of hard work and learning, I had a foundation degree in dog behaviour and training, and, the year I completed it, the college offered a two-year top-up course to turn this into a bachelor of Science (BSc) with Honours qualification, if a piece of research was carried out and written up. I signed up for this, and am very glad I did so, since I learned some interesting things about training my own dogs.

I want to talk about the research I did, and to make it easier to understand I will give a brief explanation of some of the theoretical terms I used in it – specifically what is often called the four quadrants of operant conditioning. Operant conditioning is a type of learning where behaviour is controlled by consequences. For example, we usually learn pretty quickly not to test whether or not the electric hob is on by pressing a hand on it, because if it is on, what happens next is lots of pain.

Key concepts in operant conditioning are –
• positive reinforcement
• negative reinforcement
• positive punishment
• negative punishment

This was where I used to get confused and would stop grappling with the terms, partly because there is a big difference between how these words are used in everyday English and how they are used by behavioural scientists. We know that positive means good and negative means the opposite. Yet in the terminology of behavioural science, positive in operant conditioning simply means that something is added to the situation. This is where it gets confusing:

a piece of sausage that a dog might enjoy eating (positive reinforcement to increase the behaviour), and a reprimand that a dog may try to avoid (positive punishment to reduce the behaviour) are both 'positives,' as they have been added. Negative means that something is taken away, so could include leaving the room for a few seconds after being bitten by a rambunctious puppy, as well as the release of a dog's ear from the painful clasp of a child's fingers. The first takes away your attention, and the second takes away the dog's pain.

The terms punishment and reinforcement, when used in operant conditioning, refer to the frequency of actions. If I gave my dog a sausage every time he sat, and noticed that he sat more often, I could say that his action was reinforced by the sausages. Conversely, if I walked out of the room every time my dog jumped up at me, and I noticed over time that his jumping up occurred less often, I could say that jumping up was punished by me leaving the room. Reinforced actions happen more often; punished actions happen less often.

To me, the quadrants make more sense if they are thought of as acting in pairs – with positive reinforcement and negative punishment working together and positive punishment and negative reinforcement acting as a pair in a similar way.

I would advise caution, too, when it comes to thinking about operant conditioning in general, because the terms used say nothing at all about the role that emotion plays in behaviour, or about the need to adapt, to accept that individuals learn different things at different times, and that the frustration of a student who doesn't understand what is required can make even the process of learning unpleasant.

What I chose to research was how effective different training methods are, and, to do that, I needed to do some detailed planning, and gather data, and examine this to see what conclusions could be drawn. Planning was the largest part of the work, and I thought about this for a long time before getting help from my college tutor. Eventually, I had a plan: I would use my own three dogs as a study group, and would experiment with using three different training methods to teach three different tasks.

Each task would begin with the dog sitting on a mat, placed two metres (6.56 feet) away from a target object. On cue, the dog would rise from the mat, walk to the target object and place their front paws on/in it. Task 1 object was a plastic chopping board; task 2 object was a hoop on the floor, inside which the dog had to place his front paws, and task 3 object was a door mat.

The three training methods I used were: clicker+food; clicker+food with TTouch given immediately prior to the session; simply leading the dogs to the target using a lead attached to a harness. The first two methods could be described as primarily based on positive reinforcement. Giving the dogs food would, I hoped, mean that they would be more likely to move onto the target object, whilst other actions, such as simply remaining on the start mat or wandering out of the training area, would be negatively punished, as no food

# Canine aggression

would be forthcoming. The third method could be described as mostly based on negative reinforcement, since the pressure from being guided by the lead would be removed once the dog was where I wanted him or her to be with, of course, actions like wandering off or refusing to move being positively punished by the addition of pressure from the lead.

I had chosen to use a method from the negative reinforcement/positive punishment side of operant conditioning because I had been at a conference not long before signing up for the BSc course, where a method based mostly on negative reinforcement was used for working with dogs who behaved aggressively. The speaker claimed that the method was far more effective than any based mostly on positive reinforcement. I didn't agree – and from what I saw from her practical sessions, neither did the dogs – but I was intrigued by the concept and its possible results if I used a similar method with my dogs.

Calgacus, Katie and Cuillin were my assistants in this study, and I was interested in what their input would be. Testing to examine how effective something is is tricky, since lots of things can impact on the results. Using just three dogs – each of whom would learn all three tasks – I was keen, insofar as possible, to avoid a situation whereby the dogs became used to what was required of them (placing paws on/in an object), once they realised that this was the same for all three tasks.

I took several steps to try and avoid this happening: each dog learned the three tasks in a different order; I used different target articles for each task; each task was taught in a different room; the cue to move was non-verbal. I also tried to neutralise other elements that may affect the speed at which my dogs learned, known as experimental artefacts.

To avoid hunger and frustration affecting the results I conducted the sessions at the same time each day – an hour before the dogs would normally receive their evening meal. I also made sure that I trained the dogs in the same order every time, so that any frustration they felt from having to wait their turn was around the same for each method.

Before starting I paired each training method with a task. Task 1 (placing front paws on a chopping board) was always taught using the TTouch followed by clicker method. Task 2 (placing front paws in a hoop) was always taught using the clicker method. Task 3 (placing front paws on a door mat) was always taught by guiding the dogs with a lead.

As mentioned, I was keen to ensure that each task seemed like something new to each of the dogs, and my use of a different target article for the three tasks (and their associated training methods) described, was part of this strategy. To further tackle the issue of the dogs learning the last task more quickly due to familiarity, I used a different room of the house for each task. Task 1 was always taught in the spare bedroom. Task 2 was always taught in the living room. Task 3 was always taught in my bedroom. In addition, I taught the three methods to each dog in a different order. For Calgacus, the order was Task 1, then

Task 3, and finally Task 2. For Katie, the order was Task 2, then Task 1, and finally Task 3. For Cuillin, the order was Task 3, then Task 2, and finally Task 1.

I wanted to be sure that differences in my voice couldn't affect the results. I worried that I would be more enthusiastic in giving the cue to move when I was using the training methods that I preferred. Or that if I'd had a bad day at work, frustration would creep into my voice and affect how the dogs felt about that evening's training sessions.

So I bought an electronic device that could produce different sounds, and taught those as the move cues for each task. Task 1 was a trill  sound. Task 2 was a ping sound. Task 3 was a chirp sound.

An accurate record of progress was an essential part of my research, so I wrote up the details of what happened during each session as soon as it was completed.

It wasn't an ideal setup at all. There were some artefacts that I could not entirely remove from the experiment due to various constraints, including time and funding. One of these was my own bias. Results of tests – including those that involve studying behaviour and learning – are heavily influenced by the beliefs and desires of the experimenter, not because the experimenter is deliberately cheating, as the influence is usually subconscious. Of course, as I was running the experiment, my own beliefs and previous learning about dogs would have an impact on the results of the experiment. I prefer teaching that uses the positive reinforcement/negative punishment side of operant conditioning, so I tried to neutralise my bias as far as I could. For the method that involved leading the dogs, I made sure that this would be safe and acceptable to my dogs, using a harness and lead to gently guide them onto the target object. All of them were used to wearing harnesses for walks, and used to there sometimes being a gentle tug on their lead, so I was confident that this would not upset any of them. Having this level of confidence was one of the advantages of using my own dogs for the study.

Another thing I did was ensure that Calgacus, Katie and Cuillin could choose to end the session if they wished – if any of the dogs left the room, that was the end of the session – which I hoped would also make it easier to gauge their reactions.

The final thing I did to try and prevent me subconsciously sabotaging the training was introduce a rule that none of my beloved dogs would be subjected to ongoing sessions if the training wasn't proving effective. If a dog didn't respond correctly twice in a row, the session would be brought to an end. The subsequent session would begin from one step before the end of the unsuccessful session. For any of the tasks, if a dog participated in ten training sessions without making progress, the task in question would be dropped.

I like using the clicker to teach, and had been using it successfully for years at this point. I was also a qualified TTouch practitioner with years of experience in terms of time, energy and money, spent learning about TTouch, as

# Canine aggression

well as the firm belief that it was a key factor in Calgacus learning to overcome his issues with other dogs.

I worried that I might subconsciously do things that would make the TTouch method more successful, so all training for Task 1 took place in the spare bedroom, which none of the dogs were used to going in, so would be the hardest for them to learn in.

I received approval from the college ethics board for my plan, and was happy I'd done all I could to ensure my dogs' physical and emotional well-being during the experiment.

I kept the data I gathered simple. For each session I listed the name of the dog, which of the three training methods I was using, the stage we were at on the training plan, the number of correct responses out of five repetitions of the stage of the plan, what I was going to do in the next training session, whether or not the session finished early, the date of and notes about anything extra that the dog did, and what food I used for the two methods that involved a clicker.

When I was examining the results, to gauge which training method was most effective, I looked at how many training sessions it took to get to the end of the plan with each dog. The results were – at least to me – quite interesting.

## Lead method

None of the dogs learnt Task 3, and, with each of them, I stopped training this after ten sessions with no progress. This is not to say that learning didn't occur, as they all learned that if they left the room, they didn't need to participate any longer. It took Calgacus eight sessions to begin doing this; Katie four, and Cuillin two. None of my dogs has a tendency to try and leave the area when we are learning together – not unless something interesting is going on elsewhere, at least – so this was a new experience for all of us.

This method also resulted in the dogs finding other ways to avoid what I was trying to teach. During three of his sessions, Cuillin sniffed around the room and pawed at various objects. At the start of four of the sessions, Calgacus backed away from me, and tried to avoid going into the room. During four sessions, Calgacus lay down on the start mat when I put pressure on his lead – a tactic that made guiding him impossible. Katie was the most tolerant of the three. There was just one session where she remained seated on the start mat, using her body weight to prevent me from guiding her onto the target.

I was surprised by the strength of objection that my dogs displayed to this method. I had expected that their familiarity with and usual disinterest in gentle tugs on their leads would mean they'd not be averse to this method.

And it turns out I also had an inflated view of the strength of the relationship I had with them. My dogs were used to learning new tasks at home, and I'd always thought that the bonding and challenge of this was an agreeable part of our daily life. I had fondly imagined that our shared history of enjoyable learning would mean that, even without the usual fun, games and sausages, they would gladly go along with what I was trying to teach them. That they did not was a surprise.

## Clicker–only method

These sessions pretty much progressed as I would have expected, and all of the dogs learned the task fairly quickly – Calgacus in six sessions; Katie ten, and Cuillin eleven – though each dog tried to avoid what I wanted him or her to do to some extent. Calgacus and Katie each had one session where they lay down and refused to move from the start mat, and Cuillin did that once, too, as well as leaving the room on two occasions, thereby ending the session for those days.

## Clicker and TTouch method

Combining TTouch with clicker produced more surprising results. I expected this method to be successful, and that the TTouch element would enable quicker learning, but was surprised by just how much quicker they did learn. Each dog learnt the task in three sessions – the minimum number of sessions permitted in the experiment – and no sessions ended because a dog left the room; neither did any of them refuse to move during a session.

The results were especially surprising for a couple of reasons. During Katie's final session with this method, fireworks were being let off. Katie's usual response to the sound of these was to run to the nearest window, look out and bark at them. However, she ignored them completely, focusing entirely on the task in hand. The sessions took place in a room that Cuillin is a little unsure of due to its laminate flooring, yet, for his three sessions with this method, he was quite at ease, and made no attempt to leave the room.

It appears that doing some TTouch with the dogs, prior to teaching them with the help of a clicker, enabled them to reduce their learning time by more than fifty per cent, with training going more smoothly: a good reminder of just how powerful TTouch is as a teaching method. To cut learning time in half is a significant improvement – especially given the attempts I had made to disadvantage this style of training.

Starting from a point of having little interest in understanding the theory behind dog training, I found myself conducting a study into it, and using it to critically examine my own preconceptions and habits when it comes to teaching dogs. My opinion has changed over the years, and I can see that understanding the theory is vital when it comes to working out best practice.

# 30 Health matters

I'd blamed the progesterone injection he'd had for Calgacus' attack on Monty, but had lost count of the number of dog behaviour experts and vets who'd told me that I was wrong about this, and something else had been responsible: probably his breed and him being an adolescent. However, I held firm in my belief that the change in Calgacus had been too rapid for it to be related to him growing up.

I went to a conference where one of the speakers was a retired vet, who'd specialised in behavioural problems in dogs, with a particular interest in how hormones affect behaviour. I listened intently as he spoke about how strongly physiology can impact on the ability of humans and dogs to feel pleasure, and how those feelings may affect that behaviour. I sat still, hardly breathing, unaware of the chair under me, glued to his words, my mind whirling.

During a question-and-answer session, I told him about Calgacus and the progesterone, and asked if he thought that it could have resulted in the aggression I saw. His answer was emphatic: yes; he had no doubt that it had.

He went on to explain that progesterone is a hormone that can strongly affect an individual's ability to feel pleasure. He told me that Calgacus would have been plunged into an immediate, dark depression, which he couldn't have understood or been prepared for. The likely effect, this fascinating man went on to say, was that all of those things from which Calgacus would have expected to derive pleasure – eating, meeting other dogs, being petted, walks, – would have left him feeling nothing: a huge emptiness where once there had been joy. Eventually, he said, as this continued, Calgacus would have become frustrated at the lack of pleasure in his life, resulting in him becoming snappy and short-tempered, and ultimately causing him to lash out at Monty.

I felt, in equal measure, sadness at what had happened to my poor dog and little Monty, and relief to have confirmation from an expert about how hormones impact behaviour, causing the aggression that Calgacus exhibited.

That brief conversation helped me so much, ridding me of the last of my worry that, in trying to help Calgacus get over his problems, I was making excuses for a dangerous dog who should be kept away from other dogs for the rest of his life. (The other thing it did was encourage me to always keep a close eye on the behaviour of my dogs, since this can be a strong indicator of health issues.)

Katie had been happy and well since undergoing allergy testing and a change in diet some two years previously, and was enjoying having Cuillin live with us. Then, when Cuillin was about a year old, she began to bark more frequently once more, and seemed bothered by things she'd barely noticed previously.

One lunchtime, I came home to spend time with the dogs, and let them out into the garden. On the way back in, Calgacus and Katie were shoulder-to-shoulder as they came through the doorway, and Katie turned and launched herself at Calgacus, her tawny shoulder slamming against his brindled one; her face twisting as she snarled and grabbed him by the ear, biting down hard enough to leave a small lump.

Maybe it was nothing. Maybe Calgacus pushed past her; maybe she had a bit of a sore back; maybe lots of things. Or maybe not: behavioural change like this can signal a medical problem. Dogs can't tell us when they are feeling ill, are sore, becoming confused, losing their memories. They can suffer, but can't put their suffering into words.

I wasn't sure about Katie, so made an appointment with my excellent vet, who wasn't completely able to disguise his sigh when I walked in with a happy, apparently healthy Katie, who greeted him happily and showed every sign of being utterly delighted to be there.

"I think Katie is unwell. She bit Calgacus on the ear yesterday." I explained.

"Yes – and how is Calgacus?"

"He's okay. There was no broken skin – it's just she's never bitten him before."

I was aware as I spoke that it sounded like there was no good reason for us to be there, and could understand why the vet might not be thrilled about being asked to investigate the health of such an apparently well dog.

"Would you consider seeing a behaviourist with her?" He asked – speaking slowly and with some hesitation, "she seems very healthy, and picking on other dogs is much more likely to be a behavioural problem."

"I'd like us to do some health tests first. I'm worried that something is wrong with her." I stood firm.

I cannot speak too highly about how careful the vet was in checking Katie, even though he clearly didn't agree that there was anything wrong with her. He listened to her heart; took her temperature; weighed her; took blood to test. We discovered that she'd lost a few kilos in weight, which was strange since she was eating well, and actually ate more than Calgacus every day, even though she was smaller than him.

One of the blood tests revealed signs of a digestive problem, so we then went to a local veterinary hospital for Katie to have a biopsy, which confirmed she was suffering from inflammatory bowel disease – an incurable disorder. She had no obvious signs of the illness, other than weight loss, but must have been feeling bad for some time.

The diagnosis allowed us to manage Katie's condition, and she was absolutely fine some of the time. But, as the months passed and the condition worsened, she lost more and more of herself to sudden bouts of anger.

At the same time, my wonderful parents were also going through a tough time, as Dad had been diagnosed with dementia. Although he was fine for

# Canine aggression

some time, and medication managed his condition, over time, the illness took much of his gentle, cheerful personality and left him sometimes violent, often sad, and very confused. My Mum did the most amazing job of caring for Dad, and advocating for him, and I tried to follow her lead and do the same for Katie. Mum and I would speak on the phone most days, and swap stories of how we were getting along with these tricky caring responsibilities. Going into work became a lifeline for me – a calm routine and a break from what was happening at home, where all I had to think about was solving computer problems.

Home life was tense. Whenever Katie's health took a dip, she would lash out at one of the other dogs, mostly at Calgacus, though sometimes Cuillin would bear the brunt of her anger. She never actually harmed either of them but both became frightened of her. The calm, supportive and loving household that we had enjoyed disappeared as everybody tried not to upset Katie – without knowing what *would* upset her. For a while, Katie went to stay with friends of mine for a few days every week. She loved being with them, and they had the facilities to care for her without putting the welfare of other dogs at risk. This allowed Calgacus, Cuillin and I a chance to rest and relax.

Regardless of the situation, Calgacus remained steadfast in his attempts to help Katie. Being larger than her, as well as healthier and stronger, if he had chosen at any time to react with anger he would have had the advantage, but he never did. Instead, he moved with care at home, sliding off the couch slowly to avoid startling her; waiting in doorways to receive her okay to move into the room. When – as sometimes happened– his precautions failed and Katie attacked him anyway, he would turn his shoulder to her rage and stand still, waiting quietly for me to help him; to take her away, before slowly moving away to lie down.

Some dogs would have taken advantage in these situations, grabbed the chance to retaliate while their attacker was restrained. But not Calgacus. He remained the skilled, gentle, peacekeeper – even under extreme provocation.

I appreciated his stance more than I can ever say. Thanks to him – and to Cuillin's care, too – it was never necessary to completely separate them from Katie. As long as I was with them, and we were all careful, it was possible for Katie to relax with the rest of us.

At first, another diet change – I think the recipe was made from rabbit and potato – had helped Katie feel better for a while when she put on weight and seemed more like her old self. Her behaviour deteriorated once again and we visited the vet, who prescribed medication to suppress her immune system in the hope that it would control the symptoms of her illness. This worked really well at first, but then the medication's side effects began to make her weak and unwell, and we were forced to reduce the dosage.

Time passed, and Katie seemed to be coping most of the time, although her quality of life had become limited. She was so unpredictably fearful of things that I couldn't let her meet new people on walks, and had to limit her interaction with even those she knew. We avoided other people's dogs, too, in case Katie became upset and lashed out at them. Our walks were short, and in familiar places to try and preclude the chance of coming across anything new and frightening. At home, Katie had limited access to Calgacus and Cuillin because they worried about being attacked, but Katie clearly wasn't happy about her friends avoiding her.

Eventually, her behavioural problems began to escalate – always a sure sign that she wasn't well – and neither I nor the vet wanted to give her more medication because of its side effects.

Katie's quality of life had reduced so much that I knew it was time to say goodbye. On November 26, 2010, I took her to the vet that she had loved for the whole of her life, and he helped to end her life. She was just eight years old.

I was devastated to lose Katie, but drew comfort from the knowledge that death had released her from her suffering.

Even before Katie's final battle, I became used to closely observing the dogs for tiny changes that might be something to worry about, so when, on the first day of 2010, I noticed that Calgacus was pacing more than usual, and every so often would go into the hallway and lie on the cold, laminate floor for a few minutes, I became concerned.

When I went to him and stroked him, his back was arched and he flinched – just a little, a small flicker of movement – when I touched his stomach. This was so unusual that I phoned my vet and explained to her what was happening.

"Bring him. I'll meet you there," she said.

On arrival at the clinic she told me: "I know you're level-headed so I've not phoned the nurse to come in since it is a holiday. I thought you could lend a hand." Then pointing at the man standing next to her she added: "I brought my boyfriend along, too, in case we need to lift Calgacus."

The boyfriend looked grey and hungover, but Mumbled a hello and bent to stroke Calgacus who stood patiently. Only somebody familiar with him might have spotted the tense expression on his face, and the hunched way he stood to protect his stomach. The vet bent to examine him, gently running her hands over his body.

"He does seem sore. I think we should x-ray him to be on the safe side."

The boyfriend and I held Calgacus, stroked him and distracted him from the sedative being injected into a vein in his front leg, then supported him as he became woozy and slumped to the floor. The three of us lifted him onto the table so he could be x-rayed.

The boyfriend and I waited outside – me nervous, jittery and talking too much; him quiet, ill-looking, and clearly hoping for some peace and quiet.

"I can't see much gas in his stomach." The vet said. "But there is a lot of food. It seems not to be passing through as I'd expect. I'm not surprised he's sore."

"So I've brought him down here because of indigestion?" I asked. "I'm sorry to have dragged you out for that. I thought he was really ill."

# Canine aggression

"Not exactly. This can become serious. In some dogs it can cause a torsion. I'd rather put a tube into his stomach and get the food out now than send you home and have him get worse."

Her boyfriend's complexion grew paler, and he reached out to support himself with the table ledge.

For the next half hour or so the vet worked, passing a tube, pouring water and emptying the food from Calgacus' stomach into a bowl – held sometimes by me and sometimes by her increasingly pale boyfriend. She paused at one point to top up the sedation because Calgacus, gaining some awareness of his situation, tried to jump off the table. Eventually, she x-rayed him again and felt that enough had been done to ensure he was okay.

We went home tired and a bit shaken. Calgacus had painkillers, some indigestion medicine, and instructions that he not eat dried food any longer. He recovered well from the experience, and within a couple of days was back to his usual self.

Watching for changes in how dogs act can give really good information about possible health issues, and sometimes even changes in how or whether or not dogs play with each other can indicate the same. The communication that goes on between dogs has a complexity that puts it beyond the understanding of people. We try – studying, examining and discussing – to understand them and sometimes, I hope, we get at least part of the way there. There are times, though, when I see how good dogs are at communicating between themselves  and am humbled by their skill.

I spoke one day with a friend who was upset about her youngest dog, who was going through a phantom pregnancy. The dog had been quieter than usual for some time, but that morning had grabbed one of the other dogs in the household by the thick ruff of hair around his neck, and shaken him until my friend was able to stop her. I know how upsetting and difficult conflict between dogs in the same house is to deal with, and I offered to do the only thing I could think of – go and walk with her and the dog in question.

I took Calgacus along, and we made sure he was safe by having the young dog wear a muzzle. Calgacus had met this particular dog several times: young, pretty and playful, he loved her, and had played enthusiastically with her whenever he got the chance. She was one of his friends; somebody he cared about.

That day, when the young dog ran over to greet Calgacus, he sidestepped her, deliberately turning his head and body away as he walked slowly to a tree stump, which he sniffed for a long time, apparently wanting to investigate every crevice of the bark and memorise every ring of the stump. As he stood still, his friend sniffed him. She stood next to him, and, once or twice, she inserted her nose between Calgacus and the tree stump. Calgacus responded to these overtures by turning away, bending his head and neck away from her, and resuming his sniffing in a different spot. He completely refused to engage with her, and even when we were walking together, he took care never to look toward his friend.

A few weeks later, after the phantom pregnancy had ended, we walked again, and this time when she ran to him, Calgacus bowed to her before letting her chase him around. He treated her like he had always done.

We asked ourselves how he knew? She hadn't been hostile toward him, and Calgacus hadn't witnessed any of the problems between her and the dog she lived with. Was it smell; something subtle about her body language? A coincidence? Was it he who felt ill that day and not up to playing? Was the tree stump spectacularly interesting, driving all other thoughts from his head? I just don't know.

How to deal with the issue in future was a complex question. My friend, an expert in dog behaviour herself, took time to carefully consider and research it. She was diligent, speaking to several other experts in the field. While she went through this crucial decision-making process her young dog came into season again. So she waited, nervously, to see if another phantom pregnancy would occur and, if it did, whether it resulted in the same behaviour.

Throughout this time, Calgacus greeted his younger friend with playful joy, as usual, bowing to her and encouraging her to chase him. When we met for a walk after her season had ended, he turned from her abruptly when she approached him, and refused once more to acknowledge her.

The young dog was taken to the vet, who performed a thorough examination, finding no sign of a phantom pregnancy. A few days later, though, she began producing milk; very clearly experiencing another phantom pregnancy.

I still have no idea how Calgacus knew that she wasn't well and might not be her usual, playful self.

# 31 Let's talk ...

Over the years, I've discovered that searching for a quick way to deal with aggressive behaviour in dogs often leads to a path that involves being pretty unpleasant to the dogs we care about. Even those who are usually kind and gentle teachers of dogs can find they are asked to do some horrible things if their dog behaves aggressively.

With some dogs, the behaviour that is so problematic for people is an attempt at communication: a way of letting us know that they are feeling unwell, or are unhappy, or simply need some space. Growling and snarling, showing the teeth are, on the whole, elaborate displays of implied threat and intimidation, intended to create distance between the dog and things he'd rather not be close to, and are much more effective than the dog who tries to give a wide berth to what he wants to stay away from, or who sniffs pointedly at the ground, or hides behind something. Many dogs will try these methods in the first instance, only resorting to more overt threats when passive ways of saying 'please leave me alone' have failed.

There simply is no quick-fix way to deal with behaviour problems: time, patience, and a willingness to learn are essential components, and there is a certain magic whenever we commit to using kind methods, learning how to work *with* our dog, even if the dog sometimes behaves in ways that make this difficult. We must learn what upsets and worries our dog, and find ways to help him overcome these fears, becoming an expert in our dog's body language, and skilled at helping him communicate in calmer, quieter ways. Watching for small signals and acting on them; avoiding walking in places with no escape routes; having no expectation that our dog will want to play with the dogs he comes across on walks.

We grow up with the expectation that whoever is in charge lays down the rules that everybody else follows, and it's the same if we have a dog: the dog must follow the rules without question. Inhabiting a world where instead of being an enforcer of rules, we are a helper, defender and provider of guidance is not always simple, particularly in the early stages when our dog's communications aren't well understood, how to help him isn't well known, and there are too many people asserting that they can 'cure' his aggressive behaviour in ten minutes ...

But there is a huge upside to this kind approach. Learning to listen before taking action, not forcing dogs into situations they do not like, and giving praise and encouragement instead of orders improves relationships and lives. The bond that develops when these changes are made is special beyond words, and an indescribable joy. For me, that bond is worth every moment of planning, every inconvenient ducking into a driveway to avoid a dog, every sigh of disapproval from people who would rather scare a dog into doing what they want. Nothing in the world compares to it, and learning how to do this stuff is something I wouldn't have missed for anything.

When Calgacus' problems began, one of the things I struggled with most was what I regarded as his unpredictability. We'd go out walking, and would often see other dogs, and each time Calgacus would freeze on the spot, looking at the other dog. This was predictable behaviour. What happened next, however, was not ...

Sometimes, he would turn into a striped terror, straining and barking at the end of his lead, and others he was more benign, turning away and sniffing at something. Not knowing which way he would react, was a huge source of stress for me.

I feared for Calgacus' future – and for my own. I was feeling my way in this new world, working on my feeling that Calgacus was distressed by the changes in himself, but what if I was wrong? What if his reactions were decided on a whim, and were just as unpredictable as I feared they might be? Plenty of folk with more canine experience than I thought I was wrong about his distress, sure that he he'd grown up to hate other dogs – maybe they were right? I regularly lay awake at night wondering if I was making the right choices.

What I needed, as it turned out, is what I always need when learning. Time. Time and the chance to observe Calgacus meeting other dogs again and again. After I'd had the benefit of time, I could see that when Calgacus stood still it wasn't always for the same reason. Sometimes it was very obviously a warning to another dog to stay away, as his head would be slightly higher than normal, his neck stiff, and he'd be leaning forward slightly from the shoulders – these were the barking, lead-straining occasions. Other, gentler times he would stand with his head slightly lower, his shoulders down, his muscles loose and fluid.

I noticed the difference really clearly for the first time during a walk with my friend, Lara and her dog, Leo, who sometimes had trouble understanding other dogs, which meant that his social experiences hadn't always been positive. Thanks to the patient work of Lara, her husband and their friends, Leo eventually became so good with other canines that he helped hundreds of dogs learn how to play and enjoy the company of their own kind. He really was an amazing dog.

At one point in our walk, Calgacus and Leo found themselves standing close together, off-lead. They each stood perfectly still, which scared me as I wasn't sure what might happen next. Lara and I had worried that they would get out of their depths, and react badly to each other. We called to them, hoping to get them away from each other.

That was when I saw it –the looseness in Calgacus' muscles; the lowering of his head – and knew, even before he moved, that he was inviting Leo to play. Calgacus bounced on his front legs, turned from Leo, and galloped away across the grass, glancing back, moving slowly, inviting Leo to chase him.

# Canine aggression

This experience marked my understanding of something I'd been struggling with.

Time passed and, as it did, I found myself noticing more often that other dogs seemed to understand Calgacus, too. Realising this helped me relax, and I no longer feared having a barking demon at the end of the lead.

I was walking with a friend one day when a Border Collie in the same field ran toward us.

"Shall I get in-between and head him off?" my friend asked me.

I looked at Calgacus who was standing still – head high, leaning forward, eyes fixed on the other dog. I smiled.

"Don't worry about it. I think the Collie will stay away. If we just stand still here and let the dogs chat, I think things will be fine."

We stood for a few seconds, and the Collie did, too, joining us in our stillness. Then he dipped his head, turned about, and went back to his carer. I'd learned by then to trust Calgacus and his ability to be understood by other dogs.

Something else I struggled with at the start of working with Calgacus were patterns, which I looked for in the dogs I should avoid. I knew that all Spaniels were difficult at the start, and knew, too, that un-neutered male dogs were a problem. I've spoken to people who avoid Labradors or Collies or Huskies, or large dogs or small dogs, or male dogs or female dogs. While these patterns help make the world more understandable and appear safer, I eventually came to believe that this approach is too simplistic.

It seems to me that much of the time dogs regard each other as individuals, just as we do each other. Yes, there are plenty of dogs who have developed such a strong, overruling worry about particular types of dogs that they cannot see them as anything other than worrisome. But there are also plenty of dogs who decide if another dog is a potential friend, threat or neutral presence by considering them individually, and Calgacus eventually became one of these.

An event one evening at our heelwork to music group demonstrated very clearly how Calgacus had begun to see dogs as individuals. The group contained several large, un-neutered male dogs. One particular evening, we were having coffee and chatting after training. Calgacus was with me, lying by my side with his lead on (he could be with the other dogs but still needed a little help, so couldn't be free to mix with them). The other dogs included three of the un-neutered males: Alfie and Ben – who lived with my friend, Laura – and Ghillie, who we regularly walked with. Ghillie, seeing his friend lying on the floor nearby, came over to greet Calgacus. He approached in a relaxed manner, and lowered his head toward Calgacus to sniff him. Calgacus lifted his head to reciprocate. Ghillie moved on.

Alfie watched and was next to wander over, stopping for a chin scratch from me before lowering his head to sniff Calgacus' nose. Both dogs had relaxed, gently wagging tails, and soft, loose postures. The greeting complete, Alfie wandered away.

Ben had watched these events with interest: his friends had said hello to Calgacus, and this had gone well, so he decided that he would say hello, too. Ben worried about life more than did Alfie. He didn't have the same confidence, and could sometimes anticipate that situations might go wrong. Without being able to talk to him or see inside his head to know for sure, I would guess that his decision to greet Calgacus was one of those times where he wanted to do something, but was worried that it might not go well. Ben approached Calgacus in a different manner. Moving stiffly – his shoulder and hip muscles tight so that his gait was stilted – he held himself as upright as possible, keeping his head and tail high, and pushing up onto his toes to increase his height. His tail wagged stiffly and quickly, seeming to indicate agitation rather than greeting. He lowered his head to sniff Calgacus and, when Calgacus lifted his, things became tense. Ben growled quietly and flashed his teeth, whereupon Calgacus leapt to his feet, throwing himself toward Ben, using his shoulder to push Ben away. It was a slightly dramatic way of making space but it worked. Ben went to Laura and sat next to her, keeping a close eye on Calgacus. Calgacus shook himself, sighed, lay back down on the floor and closed his eyes.

Communication between dogs is complex and, for me, one of the most useful things I learned about when trying to help Calgacus. Once I had decided to adopt a gentle approach with him, taking the time to learn about, and giving myself the opportunity to notice and celebrate small changes made all the difference to my relationship with Calgacus – and to his ability to get along with other dogs.

# 32 Helping others

Calgacus made strides that I never expected him to, going beyond liking to play with other dogs to showing concern that they get along with together, and learn to behave politely. And not just those he knew and was friends with, but ALL dogs!

In the run-up to a trip to Bath to assist on a TTouch practitioner weekend, I was contacted by Sofia, who wanted to talk about her new dog, Gold: a sweet dog with smooth, short hair and gentle eyes. Gold was wonderful with people, and was getting on well with the dogs she lived with, but tended to be rough when meeting new dogs. Spending a day with a skilled dog behaviourist had allowed Sofia to feel confident that, in Gold's case, her behaviour was due to her novice attempts to play with other dogs, rather than because she didn't like them.

"Could we let her meet Calgacus?" Sofia asked. "He's so big and calm. Gold might benefit from playing with him. She wears a muzzle when she's out on walks so she won't be able to grab him if she does go over the top."

"Of course we can." I replied "Maybe Gold should wear a coat, too? Her coat is fine and won't offer much protection – I don't want her to be hurt if Calgacus gets annoyed with her. It'll probably be fine, and he is careful, but still ..." I trailed off, as my mind again filled with images of blood and screaming dogs.

"She can wear a coat but I really think she'll be fine. She runs through thorny bushes all the time and never gets injured." Sofia assured me.

Pre-walk discussion finished, and with both of us feeling comfortable, we took Gold and Calgacus to a large field and took off their leads. Calgacus stood, sniffing a bush – being as still and boring as he knew how. Even so, once she was loose, Gold was so keen to say hello that she charged and ran into him.

Calgacus took his head out of the bush briefly to roar out a loud, annoyed warning to Gold, and then returned to the scent. He stayed there for a long time, sniffing and sniffing. Gold was so excited by having access to him that every time he moved, she charged at him. After a time – and with Calgacus moving slowly, and being ready to resume sniffing if she ran toward him – Gold was able to be calm if Calgacus stopped sniffing and walked around. I don't remember him playing with Gold then but he was, by the end of the walk, happy to be wandering around with her. He remained careful not to run on that first meeting, and Gold was able to cope.

That week, with careful introductions to many of the dogs on the course, Gold was able to run with many of them. Cuillin was about seven months old at the time, and Gold had a gentle game with him, which he showed every sign of loving. Since that walk, Sofia continued to do amazing work with Gold – using kind and gentle methods to help her cope with other dogs – and she is fantastic.

One of my happiest memories of Gold is of another trip to Bath several years after that first time.

Cuillin, not having seen his old friend for a few years, was delighted to run around with her; so delighted that, when I took off his lead, he ran past her as fast as he could, close enough that her nose brushed his coat as he passed. Gold calmly watched him run without feeling the need to chase or grab him.

In some ways Calgacus' reaction to Gold reminded me of how he was with Carrie's dog, Ghillie. Ghillie was a wonderful dog, and even now I can picture him as a tiny puppy coming along to class, where he would sleep peacefully for the entire hour, ignoring everything going on around him. He grew up into a kind, gentle dog, and was a wonderful mentor for Cuillin when he was a puppy.

Ghillie, sadly, wasn't entirely healthy, and his wasn't a long life. He suffered from pain and restricted movement in his hind legs, and Carrie worked hard for and with him to help vets diagnose and manage the condition. Ghillie had a full and happy life with her. He was a young dog when his leg problems began, at a time when he wanted to run fast, chase and wrestle with his dog friends. It was easy for Ghillie to overstretch himself in the fun of a game, and end up suffering for some time afterwards. He needed playmates who could match him for energy, but who wouldn't get carried away in the game and encourage him to overdo the playing.

Calgacus proved an excellent playmate for Ghillie. Eyes gleaming, they would chase each other across fields, Ghillie sometimes barging into Calgacus, which made him happier than anything else. I could see him laugh when it happened, and run faster. Calgacus didn't forget himself, though, and seemed to know when the game was hotting up too much. He would stop, sniff the ground, move slowly, shake himself and calm things. If we noticed Ghillie limping and needed them to stop, Calgacus would stand still if I asked him to, and wait for Ghillie to calm down. In this way, Calgacus was able to return some of the great favours that Carrie and her dogs had done for us over the years.

"He likes other dogs and he wants them to be safe." I said to myself, repeatedly over the years as I watched Calgacus change. "Trust him to be kind and try not to hold him back." It became like a mantra, and watching each occasion helped make it more real to me as Calgacus and I reinvented our life together.

One summer we turned up for a camping weekend, and got chatting to folk pitched nearby.

"He's still hard work." They told me, indicating their dog – a sweet-looking Spaniel who snoozed on his bed, looking anything but hard work – "He doesn't like large dogs, and if we aren't quick enough to restrain him, he'll charge them, and then either chase them away or knock them over if they don't run away fast enough."

"That must be a worry for you," I said feeling nothing but sympathy for their problems.

# Canine aggression

"It is. He could easily get hurt if he did it to the wrong dog. It's such a shame – he's fantastic with dogs his own size or smaller."

"I'll look out for you on walks and make sure I keep my guys out of the way." I told them. Giving them space from my large dogs was the best I could do to help, I thought, while we all relaxed and enjoyed ourselves.

On the last day of the break, my camping neighbours and I were among the last to leave. We agreed to walk around the enormous campsite and pick up any litter that might have been left.

"I'll take Calgacus and go round that way," I said pointing to the left.

"We'll take our guys and go the other way." They replied.

In this way the dogs could be loose, but we wouldn't be close enough to each other to cause problems. I concentrated on looking for litter, the pine-scented air and dark green depths of the forest providing a lovely environment.

We drew closer together as we worked our way around the site until, eventually, the Spaniel approached Calgacus. He looked worried to me, and approached slowly, stiffly, head and stumpy tail up. He was doing the approaching, granted, and if he'd been worried I might have expected him to stay away, but my experience told me that, while some dogs do stay away from things that worry them, others take a more proactive approach and want to deal with that worry directly, a perspective I can understand.

"Be nice," the Spaniel's owner spoke in a low, warning tone.

"Don't worry. Leave them be. Calgacus knows what to do," I told her.

As I spoke, Calgacus turned away to sniff a bush. He waited – a still, striped figure among the yellow of the gorse. The Springer sniffed him, and moved slowly around him, still tense, still probably expecting a problem. With smooth, unhurried grace, Calgacus lifted his head and turned away. Immediately, the Spaniel was there, in front of him, quivering and tense.

Calgacus turned, adopting his previous head-in-the-bush position. His movement had been too quick, he realised, and he sought to reassure the other dog that he wasn't a risk. In doing so, he placed himself in a position that must have been at least a little bit uncomfortable: he'd turned from a dog he didn't know, who obviously wasn't feeling friendly, and positioned himself so that he couldn't see that dog; all to allow the dog to relax in his company. To those who claim that dogs don't show empathy for others, I say spend more time watching them!

This time, Calgacus waited longer before moving, and, when he did so, he was slower, smoother – taking his time to move his head and then each limb. The Spaniel initially tensed, looked like he might rush at Calgacus again, and then sighed, shook his body, and looked at Calgacus with open eyes and forward-facing ears. When Calgacus moved again, his new Spaniel friend followed, sniffing in tandem, the Spaniel showing a great enthusiasm to be close to a new friend.

That weekend involved another moment to add to the mental list of 'moments' that I used to shore up my confidence whenever I found it lagging.

I walked Calgacus on his lead across the campsite, toward the tangled confusion of pine trees, gorse and honeysuckle that marked the edge of the campsite. We were going to lose ourselves in the heady wood and floral scents, and deep green light of the forest.

We were some distance from where some of the other dogs were playing fly ball with their carers: a game that generated lots of excitement, and much barking from dogs waiting their turn. One of the waiting dogs spotted Calgacus and I walking by, and directed his barking at us. He pulled on his lead, straining to get free so he could run.

In an elongated moment where time slowed, I saw the white plastic of the pole he was tethered to bend, folding almost in two. The lead looped over it slid over the smooth bend and off the end. Freedom. The young brown-and-white dog was a blur as he ran at us, barking as he came.

His carer shouted his name loudly, imploring him to stop his headlong rush. Calgacus stopped, stood still, and looked at the much smaller dog. He didn't bark or growl or do anything I could recognise as communication, but he undoubtedly began a dialogue I couldn't see or hope to understand with the younger dog – who halted his run. The dogs were both still, standing in bright sunshine, looking at each other, talking to one another.

I have no idea what they said, but knowing Calgacus as I did, I suspect that his side of the conversation may have been along the lines of how rude it was to run, barking, toward him like that, and how much he hoped that the younger dog would stop because he would rather not have to deal with such rudeness. With no sign I could see, the conversation ended, the Spaniel turned and ran back to his carer, and Calgacus shook himself and turned to me.

Yet another of these moments occurred during one of the many group walks I took Calgacus on. A number of dogs were there that day, and one of them – a female with flowing red hair and dainty paws called Amber – wasn't terribly comfortable around other dogs. Like the Spaniel on our camping trip, she tended to chase them off or knock them over if she was worried by them.

We set off across a huge, green field – a small group of people and dogs trekking over the expanse. Calgacus was an outlier – hanging back, sniffing bushes and keeping out of the way. He got so far behind that I worried he might become a bother to people coming up behind us. Calgacus had refused to interact with Amber and, as he approached, I could see why he'd chosen to stay so far behind us.

Amber streaked out to meet him, throwing out her legs, her hair fluttering and gleaming. She barked as she ran and stopped in Calgacus' path, standing still, neck stiff, head pointing straight at Calgacus so that she resembled an arrow. Calgacus veered to go around her, and walked next to me while I fed him sausages. All the while Amber repeatedly moved into his path, taking up her arrow-like stance, but Calgacus kept his distance, and spent lots of time sniffing the ground.

# Canine aggression

With a snarl, Amber leapt at Calgacus' face, barking. The muzzle she wore bashed into him; her eyes were narrowed and her ears pinned back. She tried again – keen to drive him away. Calgacus barked back, a deep, booming noise. He loomed, leaning over Amber, using his body to make it clear how much larger and stronger he was. That was it for him. He disengaged at that point and came to me. Amber left him alone after that, choosing to stay near her carer, who she clearly adored.

At the end of the walk, we walked together, letting the dogs be near each other on their leads as we returned to our cars.

"Thanks for the walk. Calgacus was so lovely with Amber." Her carer said. "Things don't always go so well for her with other dogs."

I smiled. "It's my pleasure. I'm really pleased we could walk together."

With each of these events – and the many others that I don't have space to include – Calgacus showed a very strong level of concern for other dogs. Beginning from the position of being a bully, expecting other dogs to play with him, whether or not they wanted to, he'd first learned to be gentler, before then becoming anxious about other dogs, and behaving aggressively toward them. Somewhere during his journey to overcome his worries, Calgacus became concerned about the well-being of other dogs – even those he didn't know. There'd been a time when I'd wondered whether leaving the house with Calgacus would ever be possible without feeling terrified that we might see a dog: I could never have imagined he would become the skilled and caring canine helper that he did.

# $\mathfrak{33}$ Older and wiser?

Dogs age so much faster than we do – and for those of us who love them, watching our beloved companions go from strong, lively youngsters to old dogs, with all that aging brings, is hard. And it can happen so quickly that the change can be devastating.

Calgacus greyed; firstly his black face and then his feet and legs, changing from rich, red and black colouring. He couldn't move as fast, or walk for as long. I hugged him more and found myself being protective of him.

We avoided the kinds of dogs that Calgacus had so often helped – those who struggled socially with their own kind – and where once I would have offered to meet up and walk with people, I now said instead "I wish Calgacus was younger. He'd have loved to hang out and help your dog. He's just too old now." I wanted to save my dear friend from the potential trauma of possibly being knocked over and bullied by a younger dog. I wanted him to always believe that he could deal with anything life threw at him.

We didn't avoid other dogs entirely, though. Calgacus had his old friends who we met up with, and he derived intense delight from making new ones.

He must have been ten or eleven when a new dog moved near us. Six months old and absolutely glorious, this was a beautiful dog with a fluffy coat, bright eyes, sweet personality, and a huge desire to befriend as many dogs as possible.

She introduced herself one day, running to us after seeing us across the park. Joy shone in every step, and such was her enthusiasm for a game that the air around her looked brighter.

Cuillin wasn't happy at the intrusion. He left my side and stormed across to meet her, his anger as visible in his every stride as was her joy. Slamming to a halt in her path, he stood so that she was forced to stop, and glared at her, eyes narrowed, mouth tight. She lay down on the ground and Cuillin stood for a moment, rigid – the muscles in his neck wood-like with tension. He stared down at the new dog for a second or two before turning and running back to me. I quietly handed Cuillin a biscuit and attached his lead to keep him with me: I could see what affect the new dog was having on Calgacus.

He stood next to me, tail wagging, muscles soft, looking at the new dog. His whole body seemed suffused with happiness. When I took off his lead, he ran to her – his limbs floppy and puppyish – and stopped just in front of her, flopping into a bow before turning and lolloping away. Her eyes gleamed and she got up to chase Calgacus around the park. They raced around after each other, sometimes stopping to swap over who did the chasing and who was chased. At times

# Canine aggression

they would bump each other as they ran and share a conspiratorial glance. Calgacus played as if he was six months old again.

Fearful that my old dog would injure himself I put him on his lead after a while, but meeting this dog remained a source of great delight for Calgacus until he died. I agreed with him and shared in his joy. To see him able to run and play with such abandon after the struggles he'd had never became mundane for me. These play sessions became regular events, and Calgacus remained always enthusiastic to see his younger friend.

We gradually stopped taking part in so many activities as well. He still played with Cuillin and some of his other friends, and we practised heelwork to music for fun in the park, and went for walks. Other than that, Calgacus rested, becoming stiller than he'd been in his youth.

It had been so long since it had happened, and Calgacus was so calm that I'd forgotten how he felt about conflict between dogs. I'd developed the attitude that now he was old, he was happy to ignore dogs who weren't getting along; that making sure everybody got along didn't matter as much to him, but I was wrong.

We walked early one morning around the park, watching a cute young Spaniel who was running around his people, a tennis ball clamped between his teeth. Periodically, he'd drop the ball, and one of his carers would pick it up and throw it for him. He ignored us – he ignored everybody in the park apart from his people and his ball.

The Spaniel was joined by a dog who was walking with somebody else – another Spaniel, slightly older – who took away the ball, and then chased the young Spaniel. The younger dog looked scared – his tail was tucked between his legs, and he ran with his hindquarters hunched under.

Calgacus saw all of this but seemed unconcerned, so I didn't give it a thought. We were out together, sharing time and relaxing: the social dynamics in the rest of the park were not our concern. Calgacus and I went into the middle of the field so that we could do some of his old tricks, since he loved doing that so much.

Afterwards, as we left the field, the lead joining us, the older Spaniel approached. I took off Calgacus' lead so that he could introduce himself to the new dog without feeling constrained. I was smiling at the other person.

"Awww – your dog is so cute. My dog loves Spaniels," I said.

The words were no sooner out of my mouth than Calgacus growled and leapt at the Spaniel, insisting that the Spaniel stay still with the force of his growl and by physically standing over the smaller dog. Faced with such strength of feeling, the Spaniel lay down and remained still.

"I'm so sorry," I said, and blushed, feeling stupid.

"It's fine. Sometimes they don't get on. No harm done," the other person said.

I left the park worrying, and over the next few weeks I examined

Calgacus, searching for signs of cognitive dysfunction or physical illness. As time passed with no sign of him being at all ill, I thought again about what had happened.

I remembered how strict Calgacus had been in his youth, the lack of tolerance he had for dogs bullying each other, and I wondered. I didn't know the Spaniel – he was probably on holiday here – so can't say for sure whether Calgacus had been overly-forceful, old age making him grumpy, maybe, or whether the older Spaniel did really scare the younger one, and had then approached Calgacus in a similarly forceful way. Or maybe Calgacus just had a moment, a flash of feeling young again, and reacted as he would back then.

Calgacus might have been old but he was loving his life. He played with the dogs he liked, he practised all his old tricks in the park with me, and he slept peacefully on the sofa as much as he wanted to. Eleven is pretty old for a Bull Mastiff, though, and, having spent the last few years watching my Dad deteriorate as the dementia took hold, I was painfully aware that lives end, and not always in the way we might wish. In the event of an emergency with Calgacus, I wanted to ensure I wouldn't be forced into making a decision suddenly and in emotional circumstances. I decided that, should Calgacus need major surgery in order to survive, I would ask the vet to euthanise him. We all make different choices in these circumstances, but having watched how both Katie and my dad suffered, I couldn't face the thought of allowing my best friend to go through the trauma they had.

Of course, I hoped I would never have to make that decision. I expected Calgacus to live on for a year or two, becoming older, greyer and sleepier. He was healthy, after all, and this was nothing more than a precaution.

# 34 **Saying goodbye**

Goodbyes are hard, and 2012 began with saying goodbye to Dad. The years of dementia ended when he developed pneumonia at Christmas 2011, and he spent the last two weeks of his life in hospital – where we felt he would be happy and comfortable since he worked for over 40 years in hospitals, and loved his work. The whole family and many friends were there throughout that time, until he died.

Then, on 19 December 2012 – a Wednesday night – death came again, this time for Calgacus.

Work over for the day, I got home and took Calgacus and Cuillin out for walks before feeding them. I remember feeling tired that night, keen to stay at home, eat, watch TV, and relax. I think I put on my pyjamas after walking the dogs and turned on the heating.

Calgacus paced the house, unable to settle. Sometimes he would lie on the cool flooring beside the front door for a few minutes, but soon would be up again and pacing. His back was slightly arched and he had a pinched look around his eyes. I had seen this before – on 1 January 2010 – and remembered what the young vet had said about how dangerous it could be. I phoned the surgery to find that they had transferred their out-of-hours cover to a veterinary hospital in a different town. I wasn't happy about transporting a dog as ill as I knew Calgacus was that distance, but it seemed I had little choice. Calgacus and I got in the van and left. The time was 9pm.

At the hospital a vet came to see us. I told him about the pain I could see in Calgacus – his hunched back, the pinched look in his eyes – and about what had happened in 2010, explaining that the signs were the same, describing what the vet had done then, and what she had said about it. The vet – a man who had never met Calgacus or I – examined him and told me that Calgacus seemed okay; that there was no reason to suspect anything worrying. He suggested that a bit of indigestion might be the cause of the discomfort.

"He'll probably be fine," he said, "take him home and bring him back if he seems worse."

The distance worried me. I didn't want Calgacus' condition to deteriorate in the van where I might not see it until it was too late. And if he was to worsen at home, I would struggle to get him back into the van, as I couldn't lift him. I told the vet this, adding: "I'm scared of being stuck at home with him and watching him die in pain."

The vet smiled at my words, and then, incredibly, laughed. He told me that there was absolutely no reason to think that Calgacus may die, but if I was worried, they could admit him and keep him in for observation. I asked that they do this and also x-ray him straight away. The vet refused this request, saying that there was no reason to do so: Calgacus didn't seem very ill and couldn't be considered a priority. He said that they would keep him under observation and get to him when they could.

The consultation had taken on nightmarish qualities, and I was dizzied by the knowledge that this man, who held my dog's life in his hands, wouldn't listen to me. Having lived with this dog for eleven years, and seeing Calgacus' symptoms before didn't appear to matter.

Close to tears, I asked the vet if he could let me know when they might x-ray Calgacus, and was told that it wasn't possible to do so: they had other patients and Calgacus wasn't ill enough to be a priority. It could be hours. They'd phone if anything changed. There was no need for me to stay.

I left at that point, feeling sick and unsure of what to do. My partner and a friend were having a drink that night close to the hospital, so I went to meet them. I spoke of my worries and they sought to reassure me, telling me that they were sure the vet knew what he was doing, and they felt sure that Calgacus would be okay. The nightmare intensified. My fear mounted by the minute and felt worse because nobody would listen to me.

I left them, driving into the darkness, feeling alone and without hope. If I couldn't be with Calgacus, I wanted to be at home with Cuillin and with my friends who knew about dogs. Nobody was physically there, of course, but when I turned on the computer and poured out my worries on Facebook, support arrived: people who understood why I was so worried, and who had supported me through Katie's long illness and increasing care needs, and Calgacus' long recovery from his dislike of other dogs. I sat in front of my computer and my friends helped me feel less alone, staying with me as I waited to hear from the vet, and the horror that was to come.

Two-and-a-half hours after I had left Calgacus at the hospital, the vet phoned to tell me that Calgacus had been x-rayed. The x-rays had not been conclusive so had been sent to a specialist vet in America, who recommended that, to be on the safe side, a tube should be passed into Calgacus' stomach to try and remove the contents – which is what I had asked him to do the moment I arrived.

Despite the passing time and advice from a specialist, the hospital vet told me he didn't want to take that course of action, as he was concerned about anaesthetising such an elderly dog, and the trauma that might be caused to Calgacus by the tubing procedure. He told me that Calgacus seemed okay, still, and recommended waiting a further hour before repeating the x-rays. He also told me that some gas was passing from the stomach to the small intestine, which he hoped would mean that the problem would resolve itself. I felt helpless – completely dependent on his opinion which I wasn't sure I could trust. I didn't know what else to do so I agreed to this course of action.

## Canine aggression

We waited.

At about 1am he rang me back to tell me that there had been a 50 per cent increase in the amount of gas in Calgacus' stomach, and that he was going to go ahead and attempt tubing. This was followed by a phone call about ten minutes later to let me know that the tubing hadn't been possible because Calgacus' stomach had twisted, meaning that the tube could not reach properly it.

By the time the vet and I were having this conversation, Calgacus had been anaesthetised for the tubing attempt, and I was thankful that he wasn't conscious and in pain. Surgery is possible for torsions, and many dogs have this and survive and are fine. The surgery is serious, however, and some dogs don't do so well.

I had planned what to do in such a situation, as described in the previous chapter. I didn't want my old friend to end his life in pain and confusion. I was glad I had made that plan because it gave me clarity on that dark night. I was left with a decision that I never wanted to make, but it was something I could do for Calgacus, which the vet's lack of empathy could not diminish. Although it meant I would not have the chance to say goodbye to him, I asked the vet to euthanise my beloved dog without waking him from the anaesthesia.

The pain following that decision is indescribable, and was closely followed by guilt, which I feel, still. My old friend died in a strange place, surrounded by people he didn't know, and I wasn't there to provide comfort and support. In the bleak moments after I knew I'd never see Calgacus again, my friends were there for me, though, and got me through the trauma.

# 35 **Life after Calgacus**

In the days after Calgacus' death, my partner stayed with me, taking time off work to do so. My sister rushed to the house as soon as she heard the news, and I spoke with Mum and my brother on the phone.

I met friends often and walked with them. I took Cuillin everywhere with me, and the friends I met with were happy to have him in their home, or to meet in dog-friendly places. The support and care that came from all around helped more than I can say.

Even so, in the year after Calgacus' death, I was immersed in a world of hurt, anger and frustration. For him to die from something that could have been treated more appropriately filled me with anger. Anger at my vet for changing its practices, anger at the out-of-hours vet who wouldn't listen to me, and most of all anger at myself for not doing more.

I complained to my vet and then left the practice, finding a new place to take Cuillin, not filled with memories of Calgacus. I complained to the emergency vet practice. When it sent me a bill for Calgacus' treatment, I phoned and told them I would not pay it. When I received a reminder, and then a demand for payment within a week, I emailed to say I would not pay, and then sent a detailed complaint to the vet's governing body. The emergency vet practice stopped asking me for money after that, and left me alone.

The centre of my world was gone: firstly Dad had gone – the incredible man who'd taught me so much about being kind to others – and now Calgacus, my closest friend, and the one who'd taught me so much about myself, about life, about dogs, and about humanity.

I felt – still feel – fundamentally changed by their loss. Some friendships fell away during that time – I was grieving and didn't have the emotional energy for everything and everyone – whilst others became stronger, and I found myself spending more time with certain people than I had before.

Cuillin went from being part of a multi-dog household to having no canine companionship, and I feel the situation must have been worse for him. He'd grown into a self-contained dog, one who mostly ignored his own kind other than Calgacus, and no other dog could measure up to his friend. Cuillin coped well with it, however, not showing any huge sign of distress although the change must have been enormous.

That first year, spent writing angry, bitter emails and letters, and sometimes not being able to move for grief, I looked for things to do that would help both of us cope. That year I became forty, and as part of celebrating that milestone, my mum and sister organised a party for me – a night of laughter and

# Canine aggression

too much food and wine. Some friends and I ran a ten kilometre race each month of the year. Cuillin ran with me to train for all of them, and I'm sure that the extra exercise was good for both of us.

I looked for other things to do that would help Cuillin adjust to his new lifestyle. Longer walks with friends: walks that would have been too long for Calgacus. I walked as often as I could with friends who have dogs to give Cuillin the chance to be around his own kind. The best thing I did for Cuillin was take him once a week for day care at the kennels Dave and his partner own, where he could spend time around other dogs.

We took trips to pubs; places filled with people who talked to Cuillin, and sometimes had biscuits for him. Some of them had nice dogs for Cuillin to hang out with. In that time of exploration and seeking, a friend and I discovered that Cuillin enjoys riding on a train. He'll happily curl up on a cushion to sleep for the journey, and seems to view busy train stations as places of interest, with a high chance of dropped food to find.

Our day-to-day of life was sadder without Calgacus, and our usual routines altered accordingly. But the beneficial changes that Calgacus brought to my life have endured, and by the end of his life I was a very different person. When I look now for mentors, teachers and co-workers, I seek out those who speak about co-operation, about making life easier for all species, and about searching for ways to give more choice and freedom to those they teach. Now that I know what to look for, I find people like this much more readily.

I understand that change – even change that is wanted – can be hard. Going through it can be an emotional process ... but that is okay. I have learned to take change slowly; to plan and minimise risk as much as I can, but also sometimes to just *do* things, even while my mind screams for me to stop and return to what I know.

I feel far less alone, too, and if I struggle, I will find people who can help. My social circle has changed dramatically. When Calgacus came to live with me I think I had one friend who lived with dogs. Nobody else in my life was particularly interested in dogs, or knew much about them. Now, among my friends I count many dog experts, and people who love the dogs in their lives.

I volunteer my skills and time to organisations that operate in ways that fit with my new thinking, and have become part of some wonderful groups of people who work so hard and in such constructive ways to make truly amazing differences in the world. One of the companies I volunteer for, both as a director and as a TTouch practitioner, does truly inspirational work bringing learning about dogs into prisons and community settings, to the benefit of both the dogs and the people involved. I became involved in this work when the young woman who founded the company asked if I would do a TTouch workshop for them. I agreed, and when I saw how she taught and interacted with people and dogs, I was profoundly impressed. Here was somebody working in a way that fitted with what I'd learned about the world through Calgacus.

Years after beginning that voluntary work, I felt that I had time and energy that I wanted to devote more directly to dogs. I had provided temporary foster care for dogs in emergencies over the years, and decided that I would volunteer to foster a dog. Cuillin was happy as an only dog with regular visits to his friends, and I was sure that he would share his home on a temporary basis without problems.

The company I already volunteer for works in partnership with several small Scottish dog rescues, so, after considering for a while, I approached one of these – a rescue that specialises in finding homes for Staffordshire Bull Terriers. I had long been impressed by the organisation's cautious approach, and the time it took to find the right homes for the dogs in its care. I specifically wanted to foster a Staffie, partly because I have some experience with bull breed dogs – even if my experience comes mostly from Bull Mastiffs.

My decision was also influenced by my experience with Staffies. Dave had had two fantastic Staffies for years, and a great many of the dogs involved with working with people in prison settings were Staffies, so I'd spent quite a lot of time with these dogs by then. They are, in my experience, happy, friendly little dogs who often need help to manage their excitement and tendency to become frustrated. They get a rough deal in the UK since there are far more of them than there are suitable homes; even so, I expected a long wait for a foster dog who could live with Cuillin, and cope with living in the middle of a town.

As it turned out, the rescue had a five-year-old female called Roxy in short term foster, and was looking for a longer term solution for her. After a couple of walks with Cuillin – who appeared to feel positively about her and was happy to be around her – I agreed to foster Roxy.

She arrived at my home about a week after I filled in the application form – a bundle of excitement and nervous energy. Roxy had never lived with another dog, and expressed her excitement at being near Cuillin by repeatedly jabbing him in his neck in attempts to get him to play. Cuillin responded by removing himself from her, and lying in the big cage in the living room – a place she was not permitted to follow him into.

The first months were tricky, especially when we discovered that this was a dog who was terrified to be left alone. She would bark constantly, found it impossible to eat food left for her, and, on one occasion, destroyed a door. In a real team effort between the rescue, my partner, and Dave, care arrangements were put in place that meant Roxy wasn't left alone at all while I worked on teaching her to be less fearful of this. Cuillin and I worked to help her learn to be peaceful and gentle with him, teaching her the skills of co-operation that Cuillin had learned from Calgacus, Katie and I when he moved in.

Several months later, we all loved her, and Cuillin in particular showed real delight in his new companion, playing with her often, and even allowing her to lick his face, which isn't something he permits other dogs to do. That Roxy very often acted in ways that reminded me of Calgacus sometimes brought me

## Canine aggression

to tears. She felt like part of the family, and I found myself lying awake at night feeling upset every time the rescue mentioned that it had somebody who might be interested in adopting her.

We asked the rescue if it would allow us to adopt Roxy, and received wholehearted approval for this. Roxy became a permanent part of the family, and we are learning about each other every day, which is at times interesting and frustrating, and sometimes not so easy. Having adopted Roxy, I volunteer with the rescue in other ways when I can, because I found it a wonderful organisation to work with.

It feels right to finish my story of Calgacus almost where it began: with a bull breed dog who needs help to learn new skills, and who has so much to offer ...

# Index

# Canine aggression